Special Publication No 41

Energy and Chemistry

The Proceedings of a Symposium organised by
The Industrial Division of The Royal Society of Chemistry
as part of the Annual Chemical Congress, 1981

University of Surrey, Guildford, April 7th—9th, 1981

Edited by
R. Thompson
Borax Research Ltd

The Royal Society of Chemistry
Burlington House, London W1V 0BN

British Library Cataloguing in Publication Data

Energy and chemistry.—(Special publication/
Royal Society of Chemistry, ISSN 0260-6291; no. 41)
1. Force and energy—Congresses
I. Royal Society of Chemistry. *Industrial Division*.
II. Series
531′.6 QC73

ISBN 0-85186-845-2

Printed in Great Britain by
Whitstable Litho Ltd., Whitstable, Kent

PREFACE

For its contribution to the first Annual Congress of the Royal Society of Chemistry, the Industrial Division chose as a theme the interrelationship of energy and chemistry in everyday life. The object was to review in widest aspect the role which chemistry plays in the generation of heat, both by combustion and radiochemically, and its efficient use. Of complementary importance are the intelligent utilisation of finite fuel resources (with competing demands as chemical feedstocks) and the conservation of heat by improved thermal insulation, as well as the social and environmental consequences of operating various exothermic and potentially polluting processes on a global scale over long periods of time.

The Organising Committee was fortunate in being able to bring together speakers distinguished in their respective energy-related fields. It was of singular coincidence that the symposium should be held at the University of Surrey, situated in Guildford and thus in the constituency of the Rt. Hon. David Howell M.P., then Secretary of State for Energy. We are grateful to him for both opening the Congress as a whole and for delivering the keynote address, which is reproduced in full as the first chapter of this publication. Thanks are due equally to all other speakers, not only for participation at the symposium but for providing camera-ready copy from which the book has been produced in an endeavour to make this unique collection of papers available as rapidly and inexpensively as possible.

Raymond Thompson
Vice-President, Industrial Division

Organising Committee

R. Thompson (Chairman), H.L. Bennister (Secretary), J.I. Bullock, W.D. Halstead, C.A. Morgan, G.F. Phillips, P.V. Youle.

CONTENTS

Energy and Chemistry

By Rt. Hon. David Howell, P.C., M.P.
FORMERLY* SECRETARY OF STATE FOR ENERGY, DEPARTMENT OF ENERGY, THAMES HOUSE SOUTH, MILLBANK, LONDON SW1P 4QJ

The Energy and the Chemical Industries share a number of common features. Both play an increasingly important role in our daily lives. Indeed it would be no exaggeration to say that both Energy and the products of the Chemical Industry are quite indispensible to modern society as a whole. I have sometimes compared the role of Energy in society with that of oxygen to the human body. If I were to extend this analogy further - I would think that there is a close parallel between chemicals and the vital trace elements in the body structure itself. Energy and Chemicals are absolutely essential for health, for human progress and for economic growth.

Energy and Chemistry also interact with each other in a very special way. Many energy production processes involve chemical reactions and industrial chemists play a very significant role

*Now Secretary of State for Transport.

in our Energy Industries - the Electricity Industry, the Coal Industry, Gas, Oil and the Nuclear Industries. There are a number of chemical processes which depend in turn on abundant energy supplies, and some chemicals - such as chlorine-are very energy intensive. So there is this close mutual inter-dependence between our two areas of work.

The common ground does not end here.
There are also certain constraints under which we both operate. The problem of long lead times, the pressure of technological change, the increasing burden of massive financial investment, and the uncertainties of future demand and public acceptability are some of the factors which we have to contend with in this increasingly difficult business environment of today.

Oil Price Revolution

Our two industries are also exposed to external influences, particularly those operating at international level. One such influence is the price of oil and other energy sources as basic commodities. There has been a succession of continuing crises and shock waves first administered in November 1973 and repeated in 1979 and again most recently last year. The reality of this oil price "revolution" has however only just begun to register with some people, with the result that there exists today an enormous lag between the shock and the response.

Step changes, such as we have experienced in oil prices, were once phenomena of rare occurrence. Today, it is only realistic to expect the price of oil to continue rising in real terms over the next two decades perhaps even to twice its present level. This process may be uneven: there will be times of slackness and times of sudden increase. But this will be the trend. The reverberations of these price increases, like ripples in a pond, are felt over a wide area. Only in this case however, the "ripples" are more like gigantic ocean waves!

Some months ago I hesitated using the term "Revolution" to describe the energy events of the last 2 years. On reflection, I am inclined to think that "revolution" might not be quite the hyperbole that some thought it was. Not only have the oil price rises been massive - but the adjustment process will involve a response of equal scale. Social habits and industrial patterns will have to change, and attitudes will need to be transformed with them. But what is almost certain is that the world in future cannot hope to enjoy cheap and plentiful supplies of energy as it was accustomed to in the 60s and early 70s.

Future planning

This then is the challenge which you in the Chemical Industry and we in Energy must face and overcome. The backcloth against which we both must plan and formulate our strategies and policies is one of increasing scarcity of oil. In our view oil is likely to remain the world's marginal fuel for many years to come.

Energy today is a highly political commodity, especially energy which is traded internationally. At the moment this is true of oil, but in years to come gas and perhaps even coal traded on the world markets might be subject to the same sort of uncertainties. Who knows? This "political" dimension is clearly important for both our sectors as we plan for the future.

I know that the Chemical Industry is keenly aware of the problems facing it in the future - not just on Energy Costs, but in international trade and technology generally. Your industry has a splendid track record of high productivity, high profitability and high growth, particularly during the 60s and early 70s. I am sure that your traditional resourcefulness will enable you to weather the storms of the current recession perhaps better than most other industries.

Key changes

From my perspective as Energy Secretary, I should like to mention certain key influences which could have a bearing on your industry's performance over the next few years.

The Chemical Industry is still a relatively young and growing industry with enormous potential in the future. The energy revolution which has hit the industrialised world may be an extremely difficult episode in your industry's history - but I am convinced that out of it will emerge a new structure which will be in condition to respond to the changed circumstances of our time. Man's material progress on this planet has been largely due to his achievements in the field of science and technology. We rely on this process of discovery and innovation continuing in the future. I see the rise in oil and energy prices, disagreeable though it certainly is, as being a spur to greater innovation and ingenuity in the chemical sector as well as in a number of leading industrial sectors.

Change, although disconcerting in the short-term, provides new demands, and therefore new opportunities for industry to meet those demands and to make profits. Without change, there cannot be growth.

The change represented by higher energy prices is all-embracing. It affects all consumers in all the places where they use energy: homes, schools, hospitals, offices, shops, factories, cars and aircraft. Because of that, the new opportunities are also widely spread and at all levels of technical sophistication.

This country will be looking for ever-improving home insulation materials, improvements on heating control systems; new technology to improve heat recovery, and heat pumps; efficient coal boilers to replace expensive oil-burners; micro-circuitry to maximise energy saving; new materials for lighter cars and aircraft; and new car and aircraft engine designs.

This revolution is being led by small businesses as well as large. At the simpler levels of better housekeeping to save energy it is recruiting entrepreneurs who have never before been in energy-related activities. It is an engine of development and growth.

Energy saving in the chemical industry

There will be great scope for energy saving in the future as chemists devote more of their research and development resources to devising new products and processes which are more energy efficient. The cold water detergent is just one development which could perhaps be the forerunner to a new generation of products, which will be "winners" because they are energy efficient. A programme for innovation based on the prime objective of reducing the energy content of products and processes could be central to the Chemical Industry's strategy for the future.

I noted a recent survey by the Council of the European Federation of Chemical Industry - which showed that energy savings of approximately 2% per annum have been achieved between 1974 (the year of the oil crisis) and 1979. Most of these savings have been made possible by the improvement of the production operations and processes, and by changing the mix of products in favour of those which are less energy intensive. It is encouraging to note that over the years there has been a reduction in liquid fuels consumption, as well as a gradual reduction in the use of solid fuels. Most of the increase during this time has been provided by gas.

Energy strategies: alternative sources

The industrialised world's Energy problem is to achieve a smooth transition from an energy economy based predominantly on oil to one which makes much greater use of alternatives - coal, nuclear, energy conservation, and in the longer-term the renewable sources. That this be achieved without supply interruption is a major task. It is one for which I, as Secretary of State for Energy, have responsibility.

The broad objective in Energy policy, as I see it, is adequacy and security of energy supplies, efficiency in use, at optimum resource cost to the nation.

Adequacy of energy we certainly have, and security of supply must surely be one of our greatest assets today. Efficiency in energy use is being pursued very vigorously both by my

Department and by the Energy Industries. It is my policy to promote efficiency and competition in the Public Energy Sector so that the consumer may benefit from better service and keener prices.

We are working hard to achieve that successful transition from an economy based on oil to one based on the non-oil sources. We have never pretended that transitional help will not be needed in this process. For example, the substitution of oil by coal in the industrial heating market is an important part of the overall programme which we envisage for coal in the longer term. The £50 million grant scheme which was announced in the Budget recently is an important first step towards achieving this goal. The proposed expansion of our Thermal Nuclear Power Programme which I announced in December 1979 is the second prong of our strategy, and the increasing importance given to Energy Conservation is the third.

R & D and marketable energy conservation

Nor are the renewables in this overlooked at all. We regard them as being vital in the longer-term. But new technologies take time to develop, as you in the Chemical Industry will appreciate only too well. Time must also be allowed for the important demonstration period in order to prove the technology and the product. In many cases the lead times are extremely long and often the commercial potential at the laboratory stage is far from clear. In such instances, it is a valid function of Energy policy to provide the sponsorship which is needed to take the technology up to the Development and perhaps even the initial Commercial Stage. However, Government-funded Research and Development must be geared to the creation of market opportunities which can then be exploited by the Energy Industries and the private sector.

The product must be marketable, because the opportunities are enormous. There is a potential British market worth well over £1 billion for converting oil boilers to coal in industry. The market for Energy Conservation technology hardware will also be very lucrative. So the list of opportunities can be multiplied.

The Government backs up its policy of ensuring that the price signals are right by a strong programme of publicity and advice, and by funding demonstration projects and contributing to research on new technologies.

Government contributions

Information is the vital key to saving energy. Our information budget has risen this year by 30 per cent in real terms. Where we find that one company is saving £1 million a year from its £7 million energy bill through staff education, improved lighting and heat recovery we give that publicity so that other firms can see the opportunities that are open to them. We encourage both the appointment of energy managers and the formation of energy manager groups where they can trade information.

The Government funds the Energy Thrift scheme which provides one-day visits by consultants and industrial research associations who can give advice on basic energy-saving improvements. By next year nearly the whole of manufacturing industry will have been covered.

The Energy Audit scheme, also run by the Government, looks at sectors of industry in much greater depth, and aims to recommend major improvements in industrial processes to bring about substantial savings.

Government contribution to energy conservation research and development is running at about £9 million this year. In industry that work is concentrated on recovery and use of waste heat, instrumentation and control, waste as a fuel, heat pumps and combined heat and power. In transport, we are sponsoring R & D on engine and transmission development and the application of micro-processor control to vehicle functions.

We contribute 25 per cent of the capital costs to projects demonstrating novel energy saving technology - like the project in the glass container industry which is saving 55 per cent in primary energy terms. The total budget for the 80 projects currently being funded is £17 million, of which £4 million comes from the Government.

Government guidelines

In each of these areas - information, research and development and demonstration, the potential savings if widely applied are enormous. But the way to bring about these savings is not to pour in billions of pounds of taxpayers' money. If these savings are to be achieved in fact it will be through the constant drive of British management to cut costs in the face of higher energy prices. Government spending could never supplant or reproduce the effects of that industry-wide pressure and effort. But we do have a duty to help industry to see its own way out of its present difficulties: to present it with vivid illustrations of the opportunities. That is a policy which we are pursuing with great vigour and with remarkable results.

I know that this audience will be well aware of the opportunities which lie ahead over the next few years. Prospects are, I believe, now improving. Government has recognised the difficulties which industry faces in the short term and is helping. I have already mentioned the boiler conversion scheme. The Chancellor, as you will know, has also provided help for the large energy users. I hope these measures will help and encourage industry. The Government wishes to bring relief as far as possible, through its industrial and trade policies, to those who are caught up in the problems of transition. But more than that, I believe that Britain will soon emerge from the world recession, and when we do I am confident that the British Chemical and Chemicals Related Industries can move from strength to strength.

Alternatives to Fossil Petrol

By A. Spinks, C.B.E., F.R.S.
CHARTER CONSOLIDATED LTD., 40 HOLBORN VIADUCT, LONDON ECIP IAJ, U.K.

Anyone reviewing alternatives to fossil petrol starts with at least three problems - its high recoverable energy content, its easy transportability, and ninety years' development of the engine. The car owner is accustomed to these benefits in the form of cheap comfortable personal transport for around 500 km without refuelling. They have transformed his life, and he will not give them up easily. It is quite unrealistic to suppose that he will vote for industrial energy rather than personal transport, i.e. for a plastic bucket rather than a litre in the tank, or for any battery car that can so far be envisaged.

The battery car has recently (1) been reviewed by a committee of the House of Lords. They concluded that it was a serious option, particularly for urban transport in the next century, but that internal combustion vehicles powered by natural or synthetic fuel would stay as a main form of transport. At present, this seems a very sensible conclusion, and I support it.

I am therefore concentrating in this short review on liquid and gaseous fuels. Solid fuels can, of course, be used under water boilers, or finely divided in diesel or turbine engines. However, solid handling problems and the substantial difficulties of engine development hardly justify consideration of solid fuel devices for personal transport or aviation.

The obvious alternatives to petrol, kerosene or diesel fuel derived from oil or coal, are shown in Table I.*

Lipids such as sunflower oil are quite good diesel fuels, and the alcohols and, of course, hydrocarbons, are suitable for use

*Tables are on pp. 28 and 29.

in other internal combustion engines, either alone or mixed with fossil petrol. Alcohols cause corrosion problems and separate from petrol mixtures if water gets in. These difficulties fall with increasing molecular weights but so of course does volatility. Gases, on the other hand, require a new distribution system and a new fuel tank. A fuel tank for hydrogen is likely to give formidable design problems. Its high inflammability, liability to leakage, and poor energy content by volume have led to suggestions that it might be transported as a metal hydride, for example $Fe\ Ti\ H_2$. Iron titanium hydride can contain more hydrogen by volume than liquid hydrogen, but only 1 or 2% as much by weight, thus more than cancelling the benefit of high energy / weight ratio.

However, other benefits of hydrogen justify consideration of its production and use, and temporary postponement of doubts about everyday containers. The most important, particularly, as a space or aviation fuel, are low pollution, low corrosion, reduced engine noise, longer engine life and the tested feasibility of cryogenic storage at least for elaborate vehicles.

The idea of using hydrogen is fairly old (3). Jules Verne suggested it as a universal energy source in 1870 (2), and J.B.S. Haldane seriously promoted it as a successor to oil and coal in the twenties, mentioning fuel cells as a possibility. Actual practical work was done between 1926 and 1935 by Erren on its controlled injection from dirigible gas containers into the fuel supply to diesel engines (3). The main purpose was to improve the balance between buoyancy and fuel load, but the fuel value was demonstrated. However, the only substantial use of hydrogen as a fuel thus far is in Apollo spacecraft, following a NASA flight of a hydrogen powered aircraft in 1956. Since 1970 the importance of hydrogen as a potential fuel has been stressed by Marchetti (4) and many others.

Apart from fuel possibilities world production of hydrogen has grown at something like 19% compound since 1940, mainly for ammonia and other chemical manufacture. This growth is bound to continue and to lead to alternative sources, and perhaps eventually to fuel use. Hydrogen is also extremely important as an intermediate in the manufacture of synthetic fuels.

At present nearly all hydrogen is manufactured by steam reforming of methane or other light hydrocarbons. The only significant alternatives are electrolysis where hydro-electric power is cheap, or coal gasification. There are large electrolysis plants at Que Que in Zimbabwe, at the Aswan Dam in Egypt and in Norway and Peru. The major coal gasification plants are at Modderfontein in South Africa, and in India. I have had an opportunity of comparing costs of hydrogen for ammonia manufacture at Que Que, Modderfontein and Billingham from very cheap Zambesi power, cheap open-cast coal and North Sea gas respectively and made the ratios 4 : 2 : 1 approximately three years ago. Other authors (e.g. 5) consider electrolytic hydrogen to be about three times as expensive as that from steam reforming of naphtha. Even then, value of the oxygen needs to be included. At Que Que it goes to a neighbouring steel works. It ought to be costed in most places at the much cheaper air separation price, thus giving a true cost for electrolytic hydrogen.

The chances of reducing electrolysis cost substantially are probably reasonable. Electrolysis occurs at a surface, and the full scale benefit of surface / volume ratio is not achieved - electrolysis is commonly a modular process. Nevertheless, much can be done in conventional cells, for example by increasing current density, raising temperature, raising pressure, and improving the activity of electrodes. There is also the possibility, as in chlorine cells, of using sulphonated Teflon membranes with film electrodes adherent to them. On balance, the cost of electrolytic hydrogen could perhaps be halved by cell improvement and cheaper power, probably nuclear or hydro. This could make electrolytic hydrogen a possible alternative to much dearer fossil petrol. It would also bring nearer the eventual alternative for all fuels and organic chemicals - the shift reaction from CO_2 and H_2 to CO and H_2O, and thence to methanol. From methanol excellent light aromatic petrol can be made by the Zeolite catalysed Mobil process (6), and some catalysts will give ethylene, propylene, ethanol, and thence any organic chemical, from methanol.

All these processes deserve continued research, but so no doubt do many suggested alternatives to electrolysis. Some of these are summarised in Table II. Before considering them, it is perhaps worth remembering that the overall efficiency of electrolysis using thermal power is perhaps 10 - 12%, including the Carnot loss and electrolytic efficiency of some 40%. If pyrolysis of water were effected at something like 2500°, the theoretical efficiency might approach 80%. However, there are formidable problems in achieving and containing this temperature, and in separating hydrogen from oxygen so as to avoid recombination on partial cooling. For this reason, most thermal schemes rely on serial reactions at lower temperatures. One popular variant involves pyrolysis of sulphuric acid to SO_2, oxygen and water at about 850°, and reoxidation of SO_2 by iodine at 90°. HI is then pyrolysed to iodine and hydrogen at 450°. The net reaction is formation of hydrogen and oxygen from water at reasonable temperatures, with regeneration of sulphuric acid and iodine. There are hundreds of paper schemes of this kind (4, 7). Some depend on formation of metal hydrides from SO_2 and water; others use ferrous chloride as an intermediate reducing agent. One of the most remarkable schemes proposes injection of steam into hot basaltic magmas which contain iron oxides. The recovered steam may contain up to 3 mole percent hydrogen. As with many other thermochemical proposals the engineering problems of this geothermal / thermochemical approach seem exceedingly formidable.

A Westinghouse scheme combines thermochemistry and electrochemistry (8). Sulphuric acid is pyrolysed to SO_2, water and oxygen, and the sulphurous acid is then electrolysed to sulphuric acid and hydrogen. The thermal efficiency is said to be 50%, and a lower ultimate cost than pure electrolysis is claimed. A pilot plant project is envisaged in 1983.

Solar energy, including production of hydrogen is receiving much attention (9, 10). The most developed schemes are photovoltaic, particularly those based on large area silicon solar cells. If the solar electricity were cheap enough production of hydrogen by advanced electrolysis would be attractive. Of course, there are two very severe problems, the relatively low intensity of solar radiation, and the fact that there is most energy in the tropics where it is least wanted at present. The equatorial solar flux is about 1 Kwm^{-2} at midday. At 10% efficiency, which seems attainable by modern photovoltaic cells, a power station yielding a peak power of 1000 Mw will require about a thousand hectares of cell surface, and, of course, its average power will be much less, and none will be available at night.

The diffuse character, intermittent availability, and unpredictability of solar power, and the probable cost of hectares of hardware, make the economics of any form of solar power uncertain, and in the phtovoltaic case has led to success mainly for small installations in remote, often desert, areas, and for purposes insensitive to cost, like space exploration. Apart from the problems already mentioned the inadequacy and cost of power storage by batteries become critical as the size of an installation increases. Partly for this reason, there is much merit in considering, as Porter (12, 13, 13A) and many others have done, whether hydrogen could be produced by solar photolysis of water. Most such work is closely analogous to photosynthesis. In photosynthesis two systems, PS I and PS II, are linked by a common redox component to transfer a single electron from OH^- to a putative proton with the normal consequence of CO_2 fixation, though some plants in some circumstances can use hydrogenases to liberate molecular hydrogen. So far, only PS I has been simulated effectively using catalysts rather than biological systems, and then at rather low activity. The electron donor, which may be EDTA or cysteine, has still to be reduced by a non-biological system analogous to PS II, with liberation of oxygen. So far as

I am aware, this has not been consistently and efficiently achieved, and it remains to be seen whether non- biological photolysis of water to hydrogen and oxygen using simulated PS I and PS II systems is feasible on a practical scale.

However, there are many alternative approaches (16), for example irradiation of semiconductor / liquid junctions. Hydrogen is liberated at the photoanode of a TiO_2 based cell, and oxygen at the cathode. Perhaps the most promising of these devices is a strontium titanate ($SrTiO_3$) cell (16). However, the overall efficiency is only about 1%. Indeed, no chemical system for solar conversion has yet approached the performance of solid-state photovoltaics.

Progress with biophotolysis *in vitro* is also limited (14) though chloroplast membranes will generate reducing power *in vitro*, and this may be coupled through ferredoxin to a bacterial hydrogenase (14). Chloroplast activity, however, decays very rapidly.

Various blue green algae will produce hydrogen under nitrogen starved conditions (17). *Clostridia* also yield hydrogen under some conditions. Theoretically they might be made more efficient producers by genetic manipulation.

Apart from electrolysis all the schemes I have mentioned for producing hydrogen are visionary rather than immediately practicable, and some others could be called hypervisionary, like one proposal to use 14 MeV neutrons from a fusion reactor to split water.

In these circumstances, it is hardly worth discussing hydrogen engines, transportability and so on. However, it is perhaps worth mentioning that the first fuel cells to come into commercial use are based on hydrogen and oxygen. These are under construction for small-town or stand-by power units of about 4 Mw. However, the hydrogen is produced by steam reforming of natural gas and is therefore fossil-based. The fuel cell and an electric motor could be a feasible engine option if a light weight hydrogen container and a cheap non-fossil hydrogen source could be devised.

Hydrogen is also suitable for use in modified engines of existing type. All these hydrogen possibilities, though real and worth expanded research, are unlikely to provide an alternative to fossil petrol this century.

Methane is somewhat more likely than hydrogen to be useful. The reasons for this are its ready production from various sorts of biomass by anaerobic digestion, its easier transportability, its suitability for use in conventional engines, and its facile conversion to methanol. Before considering this, we need to review the availability of biomass.

At present, biological fixation of CO_2 is the only renewable source of carbon compounds, and, indeed, the only way in which solar energy is usefully converted on a substantial scale. However, the recovery is low. It is often said that solar irradiation of the earth's surface provides in 5 days almost as much energy as the total estimated fossil fuel reserve of about 4×10^{13} GJ, and in one hour the world's current energy consumption. However, the amount fixed by photosynthesis annually is probably not more than about 3×10^{12} GJ, or about one tenth of the total fossil fuel reserve (18) - an average efficiency of perhaps 0.1%. Of course, this might almost be deduced from the millions of years needed to lay down modest seams of peat and coal.

However, biological energy fixation under optimal agricultural conditions may be as high as 3 to 5%. The main losses arise from the fraction of light used in photosynthesis, between 400 and 700 nm, only half of the total. Then, half of the incident light is lost by reflection, translucency and so on. The efficiency of use of absorbed photons is generally reckoned at eight quanta per molecule of CO_2 fixed, instead of the thermodynamically feasible four. This results in a quantum efficiency of about a quarter, reducing the overall photosynthetic efficiency to about a sixteenth. There is then a further carbon loss through respiration giving the optimal 3-5% referred to. In practice, a more usual range in temperate regions is 0.5 to 1.3, and in sub-tropical regions about 0.5 to 2.5%. All these figures are frequently measured and reliable.

The highest figures achievable under optimal artificial conditions, for example in algal ponds or algal bioreactors, are more doubtful. Pirt (15) considers that the accepted eight quanta theory is not supported by sound energetic data. He has designed a tubular algal bioreactor, carrying a continuous culture of Chlorella fed on 100% CO_2 and the usual inorganic salts. Under best conditions, he claims 18% fixation of visible light, above the theoretical eight quanta maximum of 12%. The 18% remains to be proved (19), but Pirt's reactor is a useful and interesting contender in the biomass stakes. The product could serve as a protein food, or as an energy source. However, before examining the ways in which transportable energy might be produced from any form of biomass we have to examine the nature of the majority of the world's biomass.

Assuming a 0.1% overall solar yield, about 200×10^9 tonnes of organic substance is fixed annually, and about half of this, or 10^{11} tonnes, is cellulose, by far the most abundant organic chemical. The main uses of this cellulose are for combustion, paper, textiles, food for ruminants, and indirectly as lignocellulose, i.e., timber. There is additionally a vast quantity of waste cellulose, and, of course, lignin. However, the logistics and cost of collection are severe deterrents to use.

Obviously cellulose as straw or wood could be burned in a steam car. However, as harvested it contains 50% of water, and when dried has only half the calorific value of oil. The ancient way of upgrading it was by pyrolysis to charcoal. This also produced methanol, acetic acid, tar oils and combustible gas. It is a wasteful process, but there are modern alternatives, mostly analogous to existing processes for coal gasification and liquefaction. To obtain combustible gas or oil from cellulose, $C_5H_{10}O_5$, evidently demands subtraction of oxygen, and this may be achieved to a useful degree by heating cellulose at say 350°C with steam, and hydrogen or carbon monoxide (20). At still higher temperatures pyrolitic char from the cellulose reacts, as coke does, with steam to produce both hydrogen and carbon monoxide, but an external source of hydrogen, as also with coal or coke processes, will produce a better C/H balance and more useful products. As an alternative to production of liquid hydrocarbons biomass gasification and shift conversion give synthesis gas, which may readily be converted to methanol. Indeed the thermal efficiency of conversion of corn stover, furfural residues, or wood residues to methanol has been reported to be above 45% (21). Methanol may be added to petrol in small amounts; however, corrosion problems increase, and access of water to the petrol will obviously cause separation problems. Alternatively 100% methanol can be used in a specially designed engine.

Another means of getting methanol from cellulose is anaerobic digestion, followed by steam reforming of the methane to synthesis gas. Alternatively the methane could be used as a gaseous fuel.

Anaerobic digestion, and fuel use of the resulting methane is, in fact, already widely established as part of sewage farm practice, and small scale digesters are coming into world wide use as a desirable small scale, or so-called alternative, technology. There is also some use of sewage-derived methane as a chemical feedstock, notably at Klipspruit near Johannesburg where cyanide for gold extraction is made by ammoxidation of methane from the municipal sewage works.

However, much further work on anaerobic digestion of cellulose and other forms of biomass is required to increase yields and reduce costs. The 'other forms' being studied include such undesirable effluents as pig farm slurries, and gasohol fermenter residues. Indeed, the two major objections to energy from biomass, cost and land area, both require use of all possible residues as energy sources whether by combustion, gasification, liquefaction, fermentation to ethanol, or digestion to methane. Otherwise, many possible schemes use more energy as agricultural input, crop collection, crop transport, and in the factory, than appears in the fuel product. I shall give data on this point when I discuss ethanol.

First, however, the problem of land area needs to be considered. It is well known that heavily populated industrialised countries could not possibly produce all their energy needs through any form of solar energy, even if they dedicated the whole of the land surface to energy crops. Thus, assuming 1% photosynthetic efficiency and 30% thermal conversion to fuel, the Netherlands could produce only about a quarter and the UK about a half of its total energy (table III). If the UK concentrated on producing methanol from cellulose, and ethanol from sugar or starch, just to replace petrol, we would need about half the land-surface. In other words there is an obvious competition between food and fuel. They are, of course, energetically and chemically interconvertible.

For this reason in Europe most thought has been given to producing energy from waste land, or crop surpluses or crop residues, or other wastes. Possibilities for waste land include such weeds as bracken, or intensive coppice cultivation - which was, of course, one of the main sources of energy in mediaeval times.

I am extremely sceptical about surpluses: they cease to be surpluses when useful and in demand. Also the quantity of a surplus is, by definition, unpredictable: how could one write a capital expenditure proposal?

The obvious crop residue in Britain is straw (6 million tons of it), but the difficulty and cost of collection have prohibited much use so far - much of it is burned on the field. Waste paper would be another possibility: again the logistics and cost of collection have prevented much use. However, as prices of oil and coal rise, we are bound to look more and more seriously at possibilities of getting energy from bracken, straw, forest debris, municipal residues, and so on. Energy farming, on the other hand, is unlikely to be a substantial activity in Britain.

The countries where it is attractive are under-populated and have a good climate, like Brazil, which is well known to be heavily committeed to energy farming. It is not alone in this: about forty countries are seriously considering energy crops, mostly to obtain sugar or starch for fermentation to ethanol. This is undoubtedly the most attractive current alternative to fossil petrol, as well as the oldest - as old as the internal combustion engine. Hartmann in Leipzig promoted it as a motor fuel in 1894, and Henry Ford I gave much thought to it as an alternative and successor to fossil petrol in the thirties. This resulted in Dearborn Conferences on what is now called gasohol, and to a major fuel alcohol plant in Kansas around 1936. The major pre-war development, however, was in the Philippines. The huge Brazilian government programme began in 1973, and the US Government initiated a substantial programme in 1978/9. There has, of course, been continuous use of ethanol as a racing fuel, and as a high octane additive virtually throughout the history of petrol.

The Brazilian programme is based mainly on the two most popular energy crops, cane and cassava. They are complimentary in that cane needs moisture and cassava stands drought. Other crops frequently discussed include maize, sweet sorghum, sugar beet, and potatoes. In general, the tropics or sub-tropics are best for productivity as well as under-population, and sugar cane is probably the best crop. However, cellulose has had some consideration as an alcohol precursor. Hydrolysis of a rather intractable solid material,whether as wood, bagasse, straw, paper, or cotton waste, is the main problem. On the whole, acid hydrolysis under pressure seems to be generally favoured, but much work has been done on cellulases, and it seems probable that enzyme hydrolysis will eventually replace acid hydrolysis, which has many disadvantages (e.g. product degradation, acid recovery, etc.) If so, the potential for use of cellulose, particularly waste cellulose, as an alcohol precursor could be considerable. However, at present the main interest is in starch or sugar as alcohol precursors, and in cellulose or other crop waste as an industrial energy source, continuing the centuries old use of bagasse as a boiler fuel.

There are so many excellent accounts of the Brazilian and related programmes (e.g. 22, 23, 24, 25, 26, 27, 28) that I shall limit myself to a consideration of problems, and particularly those affecting land area, process efficiency, and cost, and to their possible solutions.

A typical programme involves growing sugar cane, harvesting it, transporting it to the factory, crushing, fermenting the expelled juice, distilling, using the waste bagasse as fuel, and disposing of residues. Most technology seems to be relatively conventional and particularly to employ conventional batch fermenters of moderate size, conventional brewers yeast (Saccharomyces cerevisiae), and conventional stills. There are many discussions of cost. Table IV seems reasonably representative, although now two years out of date. It shows that ethanol was then two to four times as expensive as petroleum derived gasoline, and that sugar cane was probably the cheapest of the four precursors surveyed. Other authors give broadly similar ratios.

Since 1979 the gap between sugar cane, ethanol and gasoline has probably narrowed still further, but not as much as the oil price alone would suggest. The cost of oil feeds back into the cost of all materials competitive with oil, as one can readily see for coal as well as ethanol. In the end there will be an increasing dissociation, but, in my opinion, not before the end of the century. However, we can assume that a doubling of overall ethanol efficiency might close the gap between ethanol price and gasoline price. This is highly desirable for countries who have the climate, excess land, and balance of payments problems that lead them to government gasohol initiatives. For the rest of us halving of ethanol price would be a valuable retardant of oil price increases.

Halved ethanol cost is unlikely to be achieved without thorough attack at every point from crop yield to product yield. Recent remarkable achievements in plant breeding, particularly of cereals, encourage the hope that yields of sugar crops such as cane, beet or sorghum, or of starch crops like cassava or potatoes might be at least doubled as those of rice and wheat have been. However, sugar cane (Saccharum officinarum) is unknown in the wild and is presumably already much hybridized. Thus it rarely sets seed, and has a high and variable chromosome number. Also it has C4 photosynthesis, using phosphoenolpyruvate rather than the phosphoglycerate of C3 photosynthesis. This reduces depression of CO_2 assimilation by high temperature or by photorespiration. Certainly, sugar cane is among the most efficient converters of solar energy so far known. Nevertheless, new techniques of genetic manipulation, and the increasing availability of so-called genebanks, ought to give more chance of controlled advance in crop performance than empirical hybridization has done. Recombinant DNA techniques can be reinforced by other advances such as cell fusion and plant cell cloning. Soon, it ought to be possible to aim at numerous specific targets, such as increased chloroplast density, reduced photorespiration, a more favourable sugar / cellulose balance, enhanced root growth, increased temperature tolerance, nitrogen fixation, and so on.

Nitrogen fixation is perhaps a doubtful benefit. The energy for nitrogen reduction, if it occurs in root nodules rather than in a catalytic converter, has to be provided by carbohydrate, and leguminous plants are understandably rather poor carbohydrate crops. Moreover, nitrogen is already fixed in cane plantations by mixed bacterial cultures transferred to new soil with the old sets used for propagation. It would possibly be better to use heavy applications of synthetic fertilizer, the manufacture of which is now carried out at very high thermodynamic efficiency. It is sometimes objected that the high energy input involved in heavy fertilizer application is a waste of valuable energy, but it can readily be shown for many crops that there is an amplification, perhaps as much as three or four fold, of energy input in the enhanced calorific value of the crop.

A similar energy amplification can normally be demonstrated for other modern farming aids such as herbicides, insecticides, fungicides and so on, and their judicious use is a major factor in the steady growth in yields of most crops. But there are further possibilities. A grass like cane or sorghum might be expected to benefit from the hormones that cause internodal elongation, the gibberellins. ICI tests in Queensland some years ago did enhance cane sugar yield, but its effect was uneconomic. Cheapening of these fermentation products by modern biotechnology could reduce cost to a point where use on cane or other grass crops could be economic. There is also much current interest in promotion of specific plant organs such as leaf, or stem or root, by chemicals.

Taking all these possibilities together with other modern advances such as programmed irrigation, there ought surely to be a chance of increasing sugar cane yield by at least 50%. This would still be no more than 40% of the maximal efficiency to be expected of an eight quantum process. Obviously, there will be increased agricultural energy input to achieve this, and increased agricultural energy to harvest and transport the larger crop. This needs careful study because, as many authors have pointed out, the energy in the ethanol is often less than the total agricultural and industrial input. Representative figures are shown in Table V. There is a substantial gain for cane and cassava only if full use is made of all residues. However, the major energy input is in the factory not on the farm.

Given improved cultivars, intensive, chemically aided agriculture, and economically mechanised harvesting and transport, the next important stage is fermentation. There are two possibilities of improvement, an extremely simple cheap process, or a highly sophisticated very efficient one. As an example of the former,

I have met a Guatemalan biochemist who stirred chopped pieces of cane in his conventional batch fermenter. As an example of the latter, one might imagine a very large single stream continuous fermenter maintained by computer control in a steady state optimised for productivity rather than growth.

It is essential that fermentation for alcohol be improved since it suffers like all fermentations from inadequate process intensity as well as substrate cost. These are the main reasons why petrochemical alcohol has been perhaps a third the cost of fermentation alcohol.

One obvious improvement is a large increase in single stream fermenter capacity: what can be done is illustrated by the ICI protein fermenter of 1500 cubic metres capacity. As with all bulk reactors fermenter capital costs tend to fall as the capacity to the power of two thirds. A second major cost reduction could be achieved by increasing substrate and product concentration. This might be attempted by using new organisms instead of conventional yeast, and by genetic manipulation to reduce product inhibition. Genetic manipulation to concentrate the organism's effects on making alcohol rather than protoplasm would also be helpful: as already stated this should be enhanced in a computer-controlled steady state maintained by advanced sensors. The temperature of fermentation could usefully be increased to increase metabolic rate and reduce heat loss. Yet another possibility is to carry out the fermentation in a column using immobilized cells.

One could envisage perhaps a thirty to fifty percent cost reduction by means such as these.

The next major cost is alcohol isolation. Conventional distillation is expensive, but the cost would be significantly reduced by higher ethanol concentration, and by vacuum stripping, and perhaps by extraction with some such solvent as dodecanol. Solvent extraction might also avoid azeotroping to get rid of the 4% water content of distilled spirit.

Finally, nothing must be wasted. Dried residues such as bagasse must either be burned to raise steam, or hydrolysed to provide more sugar, or digested to yield methane. The stillage residue forms a most undesirable effluent, but the yeasts may be used as animal feed, and the aqueous residue or, indeed, all the stillage, digested to yield methane. Any ultimate waste might be used as fertilizer to recover part of the nitrogen.

Adding all this up, perhaps rather optimistically, I conclude that there is a fair chance of halving the cost of ethanol towards that of the current cost of gasoline. There is still the difficulty of finding adequate land, and in a competition for land between food and fuel no one could vote for fuel. There is a strong chance that this confrontation may arise quite soon in the USA between corn for food and corn for alcohol.

There is, however, no difficulty, as is probably well known, in modifying existing engines to cope with ethanol. Volkswagen have done much work on this for the Brazilian market (26). All in all the Brazilian programme seems admirable to me given the Brazilian situation, particularly lack of oil, adverse balance of payments, plentiful land, rural employment problems, and climate.

Ethanol is not the only possible fermentation option. Isopropanol, butanol etc., could be considered instead, but insufficient work has been done on these to allow judgment of their promise.

There is also the possibility of producing triglycerides by fermentation, as an alternative to palm, sunflower or linseed oil for diesel fuel. For example, Arthrobacter and Candida species are known to produce up to 70 to 85% of the bodyweight as fat. There is much scope for work on such organisms, and also on algae such as Botryococcus braunii that produce hydrocarbons in potentially large quantity. Many higher plants are hydrocarbon producers, apart from the obvious rubber producers, and Calvin (29) has given a fascinating account of several of these, and particularly of Euphorbia lathyris which seems able to produce about 2½ tons of oil a hectare. This compares quite well with the yields usually cited for ethanol, that is 3,500 litres per hectare from cane, 3,200 from wood, and 2,200 from corn or cassava (30).

Thus, there are numerous alternatives to fossil petrol. The most interesting seem to me to be ethanol, methanol, hydrogen, and methane, and hydrocarbons and lipids of biological origin. Those most likely to prove competitive as engine fuels in the near future are probably ethanol, methane and methanol. However, coal seems likely to be a cheaper source of methanol, and gasoline and diesel fuel, for some time to come than those I have described, at least in places like the USA, South Africa, Australia, India, China and the USSR with plentiful cheap coal. When nuclear power becomes our major source of energy, as I believe it ought to, then production of hydrogen and CO_2, with subsequent shift reaction to synthesis gas is a very attractive option for fuel and chemical feedstock. Solar power, or solar photolysis of water are alternative, as yet unproved, options that deserve continued study, as indeed do all the possibilities that I have described and many more, since we need to keep all the options open until the next century. Taken together these options are sufficiently attractive and sufficiently promising to ensure the continuance of private vehicles. One hopes that governments everywhere will act on this assumption and collaborate in support of the vast international research development and initial investment expenditures that will be necessary to keep us all driving comfortably around our shrinking world.

TABLE I

Alternatives to Fossil Petrol

Ethanol	Methane
Methanol	Hydrogen
Higher Alcohols	Lipids
Biological Hydrocarbons	

TABLE II

Hydrogen Sources

Electrolysis	Photolysis
Pyrolysis	Biophotolysis
Thermochemical	High Energy Neutrons
Electrochemical/Thermochemical	

TABLE III

Land for Energy (EEC)

Proportion of total land needed to provide national energy demand (12 tonnes dry wt. per hectare per year)

Ireland	0.23
France	0.69
Denmark	0.88
Italy	1.00
United Kingdom	1.90
Germany	2.27
Benelux	3.50

TABLE IV

Fermentation ethanol as fuel (US cents/US gallon)
(early 1979)

	CORN	WOOD WASTE	SUGAR CANE	CASSAVA
Raw Material	44	68	56	68
Processing cost	30	40	20	33
Capital charge	40	82	30	36
Distribution	13	13	13	13
Pump price	127	203	119	150

(gasoline 60)

(After Prebluda and Williams)

TABLE V

Energy balance of Brazilian alcohol crops

Gigajoules per hectare per yr.

	CANE	CASSAVA
INPUT		
Agriculture	-17	-17
Industrial	-45	-35
OUTPUT		
Alcohol	+78	+55
Residues	+73	+38
BALANCE	+89	+41

References

1) House of Lords, Select Committee on Science and Technology, Electric Vehicles, London, HMSO, 1980.

2) Jules Verne, The Mysterious Island, 1870.

3) D P Gregory, in Hydrogen for Energy Distribution, Institute of Gas Technology, USA, 1979, 1.

4) G DeBeni & C Marchetti, in Non-Fossil Fuels, Amer. Chem. Soc. Divn. Fuel Chem., 1972, 16, 110.

5) C J Huang, K Tang, J H Kelley & B J Berger, Hydrogen for Energy Distribution (loc. cit.), 69.

6) S L Meisel, Phil. Trans. R. Soc. Lond. A, 1981, in press.

7) J B Pangborn, Hydrogen for Energy Distribution, 307.

8) C H Farbman, loc. cit., 317.

9) J A Hanson, loc. cit., 361.

10) R A Hann, Chem. Br., 1980, 16, 474.

11) H Durand, Phil. Trans. R. Soc. Lond. A, 1980, 295, 435.

12) G Porter, Proc. Roy. Soc. Lond. A, 1978, 362, 281.

13) G Porter, Phil. Trans. R. Soc. Lond. A, 1980, 295, 471.

13a) G Porter, in Energy from Biomass, Brighton, 1980, Session V.

14) D O Hall, M W W Adams, P Morris & K K Rao, Phil. Trans. R. Soc. Lond. A, 1980, 295, 473.

15) S J Pirt, Y-K Lee, A Richmond & M W Pirt, J. Chem. Tech. Biotech., 1980, 30, 25.

16) M S Wrighton, Chem. Eng. News, 1979, Sept 3, 30.

17) W J Oswald, in Energy from Biomass (loc. cit.), Session V.

18) D O Hall, Fuel, 1978, 57, 322.

19) R Walgate, Nature, 1980, 284, 586.

20) H R Appell, I Wender & R D Miller, Amer. Chem. Soc. Fuel Divn. Reprints, 1969, 13, no. 4, 35.

21) Y K Ahn, Amer. Chem. Soc. Divn. Petroleum Chem., Honolulu, 1979, 464.

22) N K Boardman, Phil. Trans. R. Soc. Lond. A, 1980, 295, 477.

23) R C Righelato, loc. cit., 491.

24) R C Righelato, Phil. Trans. R. Soc. Lond. B, 1980, 290, 303.

25) J D Bu'Lock, in Microbiol. Tech., Current State, Future Prospects, Cambridge Univ. Press, 1979, 309.

26) W Bernhardt, in Energy from Biomass, Session VII.

27) S C Trindade, loc. cit., Session I.

28) J Coombs, loc. cit., Session III.

29) M Calvin, Gen. Elect. Centenary Symposium, Science, Invention & Social Change, Schenectady, 1978, 25.

30) H S Kohli, Finance & Development, 1980, 18.

Coal and Energy: Utilisation and Conversion

By A. D. Dainton and P. F. M. Paul
COAL RESEARCH ESTABLISHMENT, NATIONAL COAL BOARD, STOKE ORCHARD, CHELTENHAM, GLOUCESTERSHIRE, U.K.

Introduction

The rapid economic growth of the industrialised nations in the 1960's was largely the result of the availability of cheap energy, particularly crude oil. World energy consumption grew at about 5 per cent per annum in that period, accompanied by a growing dependence on oil and natural gas. Events since 1973 have highlighted the dependence on oil and have directed attention towards problems of long-term supply. It is now widely accepted that world production of oil and gas will be declining by the end of the century, so that alternatives will be required. Since coal is the most abundant fossil fuel, it will have an important role to play as a future source of energy and, to a lesser extent, of chemical feedstocks.

For the United Kingdom, the total demand for primary energy in the year 2000 is estimated in the WOCOL study[1] to be 358-448 million tons of coal equivalent (Mtce), where 1 tce is equated to 277 therms. Of this total, it is forecast that 133-179 Mtce will be supplied by coal. Recoverable UK coal reserves are estimated to be 45,000 Mt, while the corresponding figures for North Sea oil are 3700-6800 Mtce and that for gas is 2500 Mtce.

While the longer-term importance of coal is thus assured, there are problems in the short-term. The current industrial recession is having a significant effect on energy consumption and, as a result, it is likely that any growth in total UK coal consumption will now be deferred until 1985. It is ironic that this set-back should have occurred at a time when both the production and the productivity of our industry are increasing. In the medium-term the most marked growth of coal usage is expected to be in industrial steam-raising and process heat. The rise in oil prices and the expectation of impending shortage is leading to a gradual decline in the consumption of fuel oil as refinery practice is changed towards the further treatment of heavy ends to increase the yield of the more valuable light products. The role of coal will thus first be to replace oil in simple combustion applications, and to take a major share in any growth in this market. We expect our sales of coal to the industrial market, now about 11 Mt per annum, to more than treble by the end of the century.

Our largest current market is power generation, which accounts for about 80 Mt per annum. Future requirements in this field will be determined by the withdrawal of fuel oil on the one hand, but also by increases in nuclear generation, and the overall growth in the demand for electricity. Taking all three factors into account, we do not expect to see profound changes in the size of this market over the next decade or so.

In the longer term, near the end of the century, there will be a need, as oil and gas production pass their peaks, for the gasification and liquefaction of coal to provide replacement fluid fuels and chemical feedstocks. It is difficult to forecast with any precision the relative sizes of various market sectors at such long range, since they will be influenced by a number of imponderable factors such as the amount of nuclear generation at that time, and the relative prices of the various energy forms. In these circumstances, the research and development programme of the NCB has the aim, having regard to work being done elsewhere, to ensure that suitable coal utilisation technology will be available in a number of principal market sectors, so that no major options are closed to us.

The objective of this paper is to describe the main features of our programme for the future use of coal and to discuss some aspects of the chemistry involved.

The Industrial Market: Steam Raising and Process Heat

When the present recession is over, we expect our major growth area to be in supplying energy to industry, to replace oil as it is gradually withdrawn from this market and to take a major share in any overall increase in industrial energy consumption. We can already offer coal at a price which, in thermal terms, is highly competitive with oil. But to give our customers the best possible service we have a number of problems to attack. A conventional coal-fired boiler, with all its ancillaries, has a capital cost more than twice that of an oil-fired boiler of the same output. This is partly because it is possible to burn oil efficiently in a relatively small volume, so that the size and hence the cost of the boiler is smaller.

It would be unrealistic to suppose, bearing in mind the natures of liquid and solid fuels, that we could ever match the specific capital cost of oil-fired plant, but the aim of our programme is to effect reductions in plant cost which, taken together with the lower fuel cost of coal, will present a highly attractive package to the industrial customer. In close cooperation with manufacturers, we are working on a number of improvements to stoker-fired boilers, including automatic ignition, better automatic boiler control, and cheaper and easier automatic handling of coal and ash. In addition to this, we are making a major effort in the development of fluidised bed combustion.

Fluid-bed systems offer a number of attractions

- the ability to burn a wide range of coal types
- relative insensitivity to the nature and quality of ash
- eventual reduction in capital cost
- advantage in some aspects of pollution control.

A 3-phase programme of development of industrial plant was started a few years ago. In the first phase, the NCB approached a number of manufacturers, offered a package of design data and commissioned the design and construction of a number of development plants, including furnaces for drying applications, hot water boilers and steam boilers for a range of industrial uses. These plants were installed in industrial sites, generally as one in a bank of two or more, which enabled us to subject the new appliances to the demands of real industrial load variations but, at the same time, permitted us to shut down and modify the equipment as part of the development exercise. That part of the programme is almost complete and we are now well advanced in the second phase in which manufacturers, using information gained in phase I, are manufacturing, on a cost-shared basis, their production prototypes of fluid-bed appliances for a number of applications. Some of these are already under test, with more to come, and we hope that in about a year's time the series production of new plant will begin. The plants under development range from 2MW to 30MW (thermal), and an example of a 3MW steam boiler is shown in Figure 1.

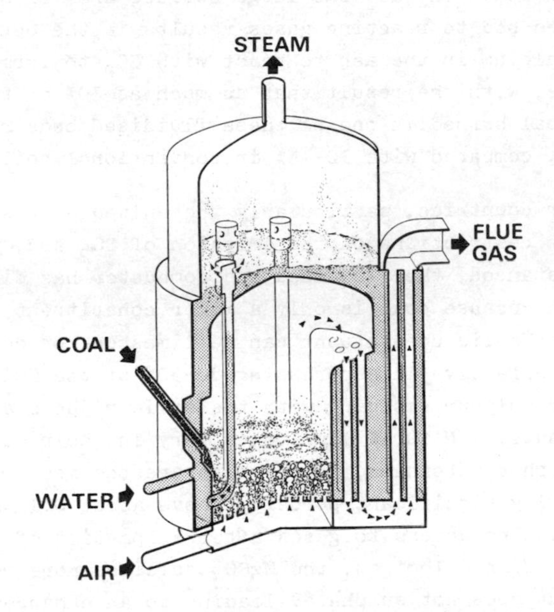

Figure 1. A 3MW vertical shell fluidised bed steam boiler

A fluidised bed is an assembly of small solid particles maintained in a state of turbulent motion by an upward stream of gas, and an important characteristic of a combustor is the relatively low fuel inventory. For example, the coal content of a bed will be less than six per cent, and in some designs less than one per cent, of the solids present, the rest being an inert material such as coal ash, sand, or other refractory. The other major characteristic is the relatively low temperature of combustion, between 800°C and 1000°C, compared with about 1400°C in conventional systems. These characteristics help in the control of pollution.

In the UK there are no restrictions on emission of SO_2, but ground level concentrations are controlled by regulations of chimney height in order to achieve the requisite dispersion. In view of the form of our legislations, the boilers and furnaces designed for the UK market have no special features for sulphur retention. In fact the large surface area presented by a fluidised bed to reacting gases results in the optimum use of any calcium in the ash to react with SO_2 to form calcium sulphate, with the result that as much as 30% of the sulphur of coal burned in one of these fluidised beds remains with the ash, compared with 10-15% in conventional boilers.

In some other countries, particularly the United States, there are stringent restrictions on the emission of SO_2 and, in these circumstances, the fluidised bed combustor has distinct attractions. Because coal is only a minor constituent of the bed, the major solid constituent can be limestone or dolomite which can be effective in suppressing 80-90% of the SO_2 emission, the sulphur emerging with the solid rejects as calcium sulphate. Natural limestones vary in their suitability for sulphur retention, those which are the most efficient absorbers having significant porosity above about 0.1 μm in diameter, allowing access to gases after sulphation of the outer layer. With dolomites, the $MgCO_3$ calcines more readily than $CaCO_3$ but does not sulphate, leading to an enhancement of the porosity. Thus the environment can be protected from the point of view of emissions to the air, but there remains the problem of the disposal of the solid residues.

In the US, studies are being made on regenerating the stone for re-use, by high temperature heating in a reducing atmosphere producing concentrated SO_2, or by hydration[2]. Other US workers are seeking applications of the reject materials: Minnick[3] has shown that, among other things, they can be used to neutralise acid mine drainage or acidic trade wastes, and that they exhibit cementitious properties for application to construction materials. Nebgen et al.[4] conclude that they are excellent materials for soil stabilisation; despite a content of only about 30% of free lime they can be used as a lime substitute on a weight for weight basis.

In the UK industrial market, where we are introducing fluidised bed combustion without limestone addition, there is the problem of disposal of non-sulphated ash. For several years we anticipate that it can readily be used as common fill but, since the production is likely to grow, there is a possible field for the inorganic chemist in finding uses for this material. Fluid-bed ashes, because they have experienced lower temperatures than those from pulverised coal in power stations, are relatively porous, non-sintered, and more easily leachable, and they may have a potential for the extraction of metals, particularly aluminium.

Power Generation

Power generation is the largest UK market for coal, currently using about 88 million tons per year, or 70% of all our coal consumption. The tonnage used for electricity is likely to remain much the same up to the end of the century but the general rise in energy prices, and the need to replace plant as it wears out, will encourage the introduction of more efficient stations. Fluidised combustion may also have applications here, particularly when operated at high pressure. The high heat transfer rates in fluidised beds allow the larger heat release rate of pressurised combustion to be matched by an appropriate heat transfer surface in a vessel of modest dimensions. Steam is raised in tubes immersed in the bed, just as in an ambient-pressure fluidised bed, but the high pressure off-gases can also be used to drive a gas-turbine.

The relatively low temperature of combustion ensures that the ash is soft and friable so that, after a hot gas cleaning stage, the residual finest ash particles are unlikely to damage the turbine.

Sulphur retention can be achieved by the use of dolomite as a sorbent. In some ways retention is more efficient in pressurised systems, since, for a given rate of heat release, gas velocities are lower and gas residence times longer in the bed. Pressurised systems also have advantages for NO_x control, if this should be required. In pulverised coal combustion, with its high temperature flames, about half the NO_x emitted may come from nitrogen in the combustion air. In fluidised combustion, with its much lower temperatures, the NO_x is almost entirely formed from the nitrogen present in chemical combination in the coal substance. Only a fraction of the nitrogen appears as NO_x in the flue gas and it appears that there are various competing reactions which produce other nitrogen compounds. The NO_x emission from a fluidised bed is dependent on the operating conditions. The mechanisms are not well understood, but higher gas pressures and greater bed depths are factors which favour the suppression of NO_x[5].

Thus, the use of pressurised fluidised beds in combined cycles is claimed to have advantages in cost and efficiency over a pulverised coal boiler with stack gas scrubbing for control of sulphur emission. The system is being tested on the 80MW (thermal) scale at a large experimental facility at Grimethorpe in Yorkshire, funded by the governments of West Germany, the US and the UK. Hot commissioning was completed late last year, and the first series of coal-burning trials has begun.

Gasification for Power Generation

There are limitations on the efficiency increase to be obtained from fluidised bed combustors since, at temperatures above 1000°C, there is an increasing risk of the softening, agglomeration and subsequent defluidisation of the ash particles. But gas turbines already work at these temperatures and future development is aimed at the production of machines with inlet temperatures of 1300°C or higher. Engines with such a potential for high efficiency would also be highly sensitive to fuel impurities which might cause erosion, deposition, or corrosion, and the only possible route for coal as the prime fuel is the production of a clean fuel gas, which need not have a high calorific value since it is to be burnt immediately and not transported to remote users.

The feasibility of coal gasification for combined-cycle operation was demonstrated at Luenen in Germany in a test programme which ended in 1977. The plant had a design output of 170MW(e) and employed five oxygen-blown fixed-bed Lurgi gasifiers producing a medium CV gas. The fuel gas, after cleaning, was used to fire a gas turbine from which the exhaust gases were used for steam raising. A new US project at Cool Water, California, is to operate a Texaco oxygen-blown entrainment gasifier to fuel a combined cycle producing 100MW(e).

The choice of a gasifier for power generation will depend upon the nature of the coals to be converted and the requirements of the electrical system. Bituminous coals, for example, soften at temperatures around 450°C and there is a tendency for discrete particles to agglomerate, a property which is exploited in coke manufacture but which can create difficulties in gasification by restricting the flow of reactant gases and reducing the surface area of the solid. As temperatures rise above 1000°C, the mineral components of the coal will exhibit softening and fusion. Properties like these can vary widely from coal to coal and in a particular type of gasifier it may be necessary to restrict the range of coals which can be converted.

Although the UK is geographically small, its coal reserves are made up of a wide range of bituminous coals, and there are similarly broad variations in the ash content, ash composition, and so on. Moreover, the proportion of fine particles, less than 5 mm, may vary over short periods between 10% and 50%. Modern power stations are large units, each drawing its feedstock from a number of mines, and the economics of scale would dictate that this would continue in the future. The type of gasifier for combined cycle operation, then, would preferably be insensitive to coal type and to variations in time of the feedstock properties.

The operational requirements for plant in the UK are also stringent. Current developments in the United States on gasification combined cycles call for base load operation or some limited load-following capability. New fossil-fuel plant in the UK must be designed to cover the range of operating regimes required to be met during its life, and these regimes will vary as, among other things, the proportion of nuclear plant increases. Thus new coal-fired plant will be required to achieve continuous operation over a range extending down to 50% of maximum load, and must also be capable of frequent overnight shut-down.

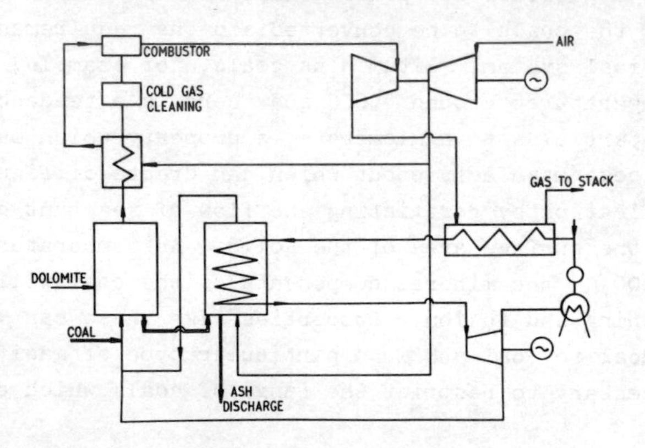

Figure 2. The NCB fluidised bed gasification process

With these requirements in mind, the NCB, with financial aid from the ECSC, is working on a gasification process, illustrated schematically in Figure 2. Coal is partially gasified in a fluidised bed, the unreacted char being transferred to a second fluidised bed where it is burned. The off-gases of the combustor, which is air-blown, provide the fluidising gases for the gasifier. The gasifier is operated in a temperature range around 1000°C, the lower end of the range determined by the need to avoid tar formation and the upper end by the requirement to avoid ash softening and consequent agglomeration. The concept is seen as being operated at pressure, but atmospheric operation of both gasifier and combustor is possible. The calorific value of the fuel gas is about 4 MJ/m^3.

The efficiency of the system would be maximised if the raw product gas could be cleaned to the requirements of the turbine without loss of sensible heat. While there are prospects for the reduction of sulphur and particulates at high temperatures, it is not possible to achieve the required low levels of alkali compounds volatilised from the ash. Some form of scrubbing is therefore necessary. The strict requirement of the fuel-gas specification may be illustrated by the statement that if the alkali concentrations are sufficiently high to be measured by existing techniques then they are unacceptable!

Sulphur might be removed after gasification by one of the existing processes, but we have the option of removing most of it in the gasifier which operates at about 1000°C, at which CaS is stable. If limestone or (for pressurised operation) dolomite is added to the gasifier bed, the solid material withdrawn from the bed will contain CaS along with some free lime and ungasified char which still contains some sulphur. In the combustor $CaSO_4$ is formed and emerges with the ash.

Preliminary experimental work at atmospheric pressure indicates that, with a 2:1 molar ratio in the feedstock to the gasifier, we can achieve an overall sulphur retention - as

$CaSO_4$ in the combustor ash - of 80-95%, depending on the temperature of the combustor, but we do not yet know the effects of operation at high pressure. The options of regeneration or disposal of the stone are the same as with fluidised bed combustors.

Stack Emissions

In addition to the gaseous products which leave the stack, mostly oxides of carbon, sulphur and nitrogen, coal fired boilers emit small quantities of particulate material. Modern particulate-control devices operate at over 99%, but some very small particles escape. These originate in the mineral matter of the coal, but the distribution of elements in these fine particles is not the same as that in the furnace bottom-ash. A relative enrichment of some elements occurs in the smallest suspended particles in flue-gas as a result of the surface condensation of volatile elements and compounds. This phenomenon is reviewed by Lim[6] and the topic of power station emissions is treated elsewhere in this Symposium[7]. It is sufficient here to remark that the concept of coal gasification for combined-cycle generation requires such a clean gas for the turbine that emissions from the stack would be correspondingly reduced.

Premium liquid fuels and petrochemicals feedstocks

Ten years ago, in an expanding economy, extrapolation of growth rates to the end of the century and forecasts of future demands for transport fuels and petrochemicals would have been accepted almost without question. Now, the situation is very different. However, amidst all the uncertainties, there are some secure bases on which to construct a strategy.

Resources of natural gas on the North Sea sector, and elsewhere on the UK continental shelf, will probably sustain our revised demand into the twenty first century[8]. Self-sufficiency in oil, however, is like to end earlier. As these resources become depleted, one option will be to revert to importing oil and liquefied natural gas, but a return to dependence on imports would require Britain to compete with other consumers when supplies will have acquired scarcity value and would be unacceptable as a long-term solution.

Scarcity will encourage substitution—a much more attractive option. The first step will be an expansion of the introduction of advanced refinery processing to upgrade fuel oil to premium liquid fuels and petrochemical feedstocks. Substitution of fuel oil by coal, either for power generation or process heat, will make this possible and so accommodate the expected limited growth in demand for these products in the mid-term. In the longer term, when this expedient can no longer compensate for the demand growth in premium sectors where no acceptable substitutes exist, coal liquefaction to make equivalent products can be introduced on an increasing scale. The National Coal Board's aim is to ensure that the best technology for liquefaction is available to provide that option.

Liquefaction technology

The routes from coals to premium fuels and chemicals have many

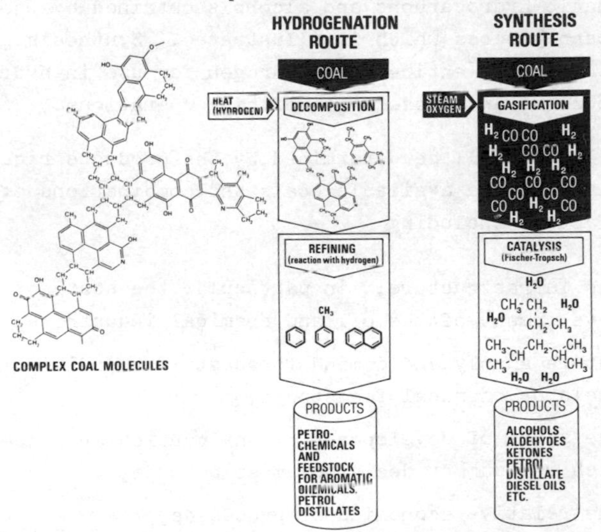

Figure 3. The hydrogenation and synthesis routes for coal liquefaction

features in common. The conversion is brought about by the application of heat (pyrolysis), by the use of solvents and by reactions with oxygen, steam or hydrogen. These, either alone or in combination, constitute the processing technologies.

The two broad divisions of the technology (Figure 3) involve either hydrogenation or synthesis. Hydrogenation involves the combined effect of heat to break down the complex molecular structures in coals to much simpler, but still ring structures. Here benzene, toluene and naphthalene represent possible ultimate chemical products.

The synthesis route involves a much more drastic breakdown of the coal structures by total gasification with steam and oxygen, the intermediate product being a synthesis gas consisting mainly of hydrogen and carbon monoxide. By suitable choice of catalysts and conditions, the relative concentrations of hydrogen and carbon monoxide can be adjusted to suit the final synthesis process which can produce, for example, methane, methanol, ammonia or the complex mixture of mainly aliphatic hydrocarbons and alcohols obtained by the Fischer-Tropsch process which is illustrated. Synthesis gas can also be converted entirely to hydrogen for use in hydrogenation processes and subsequent refining operations.

The route selected will be determined by the products required, the suitability of the available coals and combinations of many other factors including

a) the infrastructure; in particular the state of development of the oil and chemical industries,

b) future supply and demand forecasts for individual fuels or chemical feedstocks,

c) the state of development of any particular process when investment decisions must be made,

d) the relative economics of processes,

e) their relative thermal efficiencies,

f) a number of environmental considerations and

g) political or strategic factors.

Considerations such as these account for the commercial-scale operation of the Fischer-Tropsch synthesis processes in South Africa which will soon be converting over 30 million tonnes of coal per year to produce almost half of that country's demand for transport fuels[9].

These same factors have determined that -

(i) of the current world synthetic ammonia capacity of about 260,000 tonnes per day, only 5,000 tonnes per day is produced from coal while,

(ii) for methanol, the contribution is even less significant - a mere 90 tonnes per day based on coal compared with the world installed capacity of about 42,000 tonnes per day.

Nevertheless, the implication is clear. The return to coal as the source of liquid fuels and key chemical intermediates has begun overseas. The only uncertainties relate to whem the transition will begin in other countries, including Britain, and which technologies will be dominant when it occurs.

Gasification and synthesis. The original forms of liquefaction by hydrogenation and synthesis were developed and operated in Germany as complementary processes: hydrogenation mainly for the production of aviation gasoline, synthesis mainly for the production of diesel fuel and detergent feedstocks. When South Africa took up liquefaction in the fifties the choice of synthesis was determined by the high inherent ash content of South African coals which rendered them unsuitable for the existing hydrogenation process. Developments in synthesis and in petroleum refining processes, including alkylation and isomerisation, had made it possible to produce satisfactory motor spirit from the Fischer-Tropsch products.

Compared to natural gas, liquefied petroleum gas and petroleum naphtha, coal has several disadvantages as a raw material for the production of synthesis gas. Obviously a solid is less convenient to handle. Moreover, it is not of uniform size, so crushing and grading may be required. Then, whereas the gaseous and liquid feedstocks can be purified before being reformed to synthesis gas, especially to reduce their inventory of sulphur-containing components, coal enters the gasification system accompanied by mineral matter, organic and inorganic sulphur compounds, nitrogen compounds and water (Figure 4).

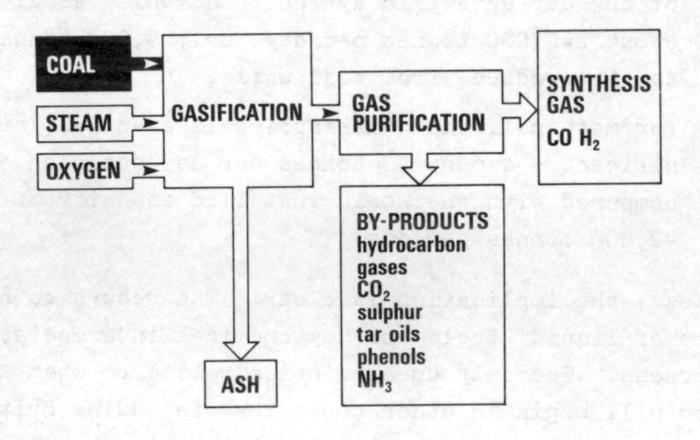

Figure 4. The production of synthesis gas from coal

With coal, the key reaction is that with steam and the measure of efficiency relates to the amount of steam decomposed, for that determines the amount of hydrogen in the raw synthesis gas. This reaction is endothermic, the heat required being generated by partial combustion. The gas composition is largely determined by the operating temperature: about 1,000°C, carbon dioxide is converted to carbon monoxide; at higher temperatures, methane cannot survive. Coal rank has relatively little significance regarding gas composition.

There are currently three gasification systems in commercial synthesis operations: Lurgi gasifiers for Fischer-Tropsch, Koppers-Totzek for both ammonia and methanol and Winkler for ammonia only. Each is representative of one of the three main gasification systems: fixed-bed, entrained and fluidised bed. Of these three commercial gasifiers, only the Lurgi operates at elevated pressure.

Improved second and third generation versions of these three gasifiers are being developed. Collaboration by Lurgi and British Gas Corporation has led to an improved gasifier which can process quite strongly caking coals and also operate at higher temperature[10]. Higher rates of gasification result and the mineral matter is discharged as molten slag. Pressurised versions of the Koppers-Totzek and Winkler gasifiers are also at the pilot plant stage. Shell is collaborating with Koppers in developing a pressurised version of the Koppers-Totzek gasifier which incorporates features of their oil gasification process. Texaco are developing their oil gasification process to handle coal. Because of the wide choice of processes available, the production of synthesis gas does not figure in the NCB's R & D programme.

All the gasifiers, no matter the rank of coal, produce medium calorific value gas. But the composition of the gas is characteristic of the gasifier (Table 1). Temperature regimes within the gasifier have greater consequences than the operating pressure. Increasing temperature through Lurgi, Winkler and Koppers-Totzek is reflected by the decline in carbon dioxide and methane contents. The higher base temperature of the slagging version of the Lurgi gasifier results in a lower carbon dioxide content compared with the conventional Lurgi but, because both are counter-current gasifiers involving carbonisation before gasification, the methane content is not drastically reduced.

GASIFIER	FUEL	CO	H_2	CH_4	CO_2	N_2	C.V.*
Lurgi (dry bottom)	Bituminous	25	40	9	25	1	297
	Lignite	20	39	10	29	<1	288
Koppers Totzek	Bituminous	55	32	0·1	11	2	277
	Lignite	56	29	0·1	12	2	272
Winkler	Bituminous	35	42	3	19	1	276
	Lignite	46	36	2	14	1	281
Lurgi (slagging)	Bituminous	60	28	8	2·9	1	360
Shell - Koppers	Bituminous	65	26	<0·4	1	n.a.	294
Texaco	Bituminous	50	35	0·1	17	1	272
Winkler (high temp)	Lignite	52	35	3	9	<1	307

* Calorific value: Btu/scf

Table 1 Typical raw gas compositions from various coal gasification processes

The main advantage of the Shell-Koppers process over the original Koppers-Totzek process is that operation at 30 bar, or higher, avoids downstream compression of synthesis gas and, incidentally, reduces the size of gas purification equipment. Improved oxygen utilisation is also achieved and this is reflected in the lower carbon dioxide content of the gas. Although the Shell-Koppers and Texaco gasifiers operate at similar temperatures, Texaco gas has a considerable carbon dioxide content. This is a consequence of using excess oxygen to meet the latent heat demand to evaporate the excess water introduced by employing slurry feeding of the coal and also accounts for several percentage points lower thermal efficiency.

However, the most important consideration is the suitability of the gas for the synthesis process. Although gas composition within the synthesis 'loop' may be quite different, fresh synthesis gas entering the 'loop' is usually close to the stoichiometric ratio. Fortunately, it is possible to adjust the hydrogen:carbon monoxide ratio by means of the water-gas shift reaction. Almost complete conversion of carbon monoxide to hydrogen is possible. The presence of methane in the raw synthesis gas is clearly advantageous when substitute natural gas is the end product. Otherwise it may have to be reformed to produce more hydrogen and carbon monoxide. Carbon dioxide can be used to synthesise methanol, but involves a higher hydrogen consumption. Apart from that, carbon dioxide and other components such as hydrogen sulphide, carbonyl sulphide and mercaptans are undesirable in the synthesis processes and have to be removed in gas purification operations. These are the three principal unit processes involved in converting raw gas to the final synthesis gas.

The Lurgi conventional gasifier was selected for the original Fischer-Tropsch plant at Sasolburg in the Orange Free State because it operated in the useful pressure range and gave a raw gas with a hydrogen:carbon monoxide ratio near that required. There was a disadvantage in that the methane formed in the gasifier and in the synthesis had to be reformed but, by establishing a neighbouring market for the plant tail gas containing about 30 per cent methane, the requirement to reform was reduced.

The SASOL organisation has developed the conventional Lurgi gasifier, in collaboration with Lurgi, to units capable of gasifying 1000 and even 1500 tonnes coal per day. Because of this familiarity with the process technology, SASOL have founded their two much larger new plants (SASOL 2 and SASOL 3) on Lurgi gasification.

The original Sasolburg plant operates two versions of Fischer-Tropsch synthesis - Arge and Synthol (Figure 5). Arge synthesis makes use of an iron/cobalt catalyst packed in tubular converters which are water-cooled. The Arge products

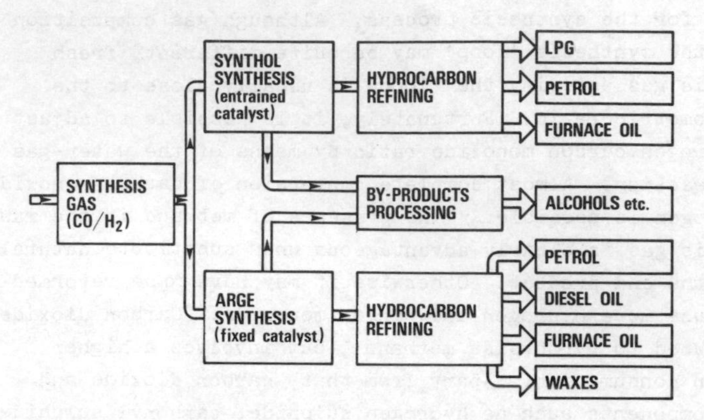

Figure 5 The Fischer-Tropsch processes employed at SASOL 1

are predominantly aliphatic in character and contain a fair proportion of high-molecular-weight waxes which make a considerable contribution to the economics of the SASOL operations. The Synthol process makes use of an entrained iron catalyst which is conveyed around the synthesis 'loop'. Synthol products are much lighter, contain olefins and a small proportion of aromatics. Although originally optimised for the production of petrol, the Synthol products can also be converted to diesel fuel.

When the decision to expand Fischer-Tropsch operations in South Africa was taken, the Synthol process was adopted because of its inherent flexibility and greater ease of scale-up. An additional reason was that, although the ARGE process yields a superior diesel fuel, SYNTHOL can yield diesel fuel of adequate quality for existing diesel engines, equivalent to the diesel fuel produced by advanced oil refineries in America.

The first of the new plants, SASOL 2, is now fully operational. Its battery of thirty-six Lurgi gasifiers will gasify about 12½ million tonnes of coal per year. The plant will produce about 1½ million tonnes of transport fuel. In addition it will produce about 150,000 tonnes of ethylene per year, equivalent to the output of a conventional naphtha cracking plant. By 1984, the sister plant, SASOL 3, will be fully operational and Fischer-Tropsch processes will be supplying about 50 per cent of South Africa's transport fuels.

All three SASOL plants have the disadvantages of Fischer-Tropsch synthesis - wide product spectrums and low overall thermal efficiencies. Market structures in South Africa can accommodate the products of these plants, but this would not necessarily apply elsewhere. The addition of the small gas grid to the original plant permits the attaining of a thermal efficiency of 56 per cent. Otherwise it would be identical to that of the new plants - about 36 per cent which is about the same as a modern power station.

Methanol and the Mobil MTG process. Development work on synthesis processes is focussed on identifying more highly specific catalysts, permitting the selective production of premium products and, incidentally, leading to higher overall process efficiencies.

The established low-pressure synthesis of methanol is one instance where high selectivity is combined with reasonable thermal efficiency - about 60 per cent. Methanol is potentially a significant motor fuel, having the advantage of a relatively high octane rating (RON 110). It has several disadvantages, however, particularly its lower calorific value and higher heat of vaporisation compared to gasoline, and it is also toxic. Methanol can be converted to methyl tertiary butyl ether, provided there is a suitable source of butenes, and this ether has all the qualities of methyl alcohol as an octane improver without its disadvantages.

Alternatively, methanol and related substances can be converted to a high octane hydrocarbon gasoline (RON 93) by a novel process recently developed by Mobil Oil[11]. The conversion - involving the dehydration of methanol to form the ether, followed by the successive formation of olefins or carbonium ions which, in turn, are converted to mixed hydrocarbons with a high concentration of aromatics - is accomplished by a highly-selective zeolite catalyst (Figure 6).

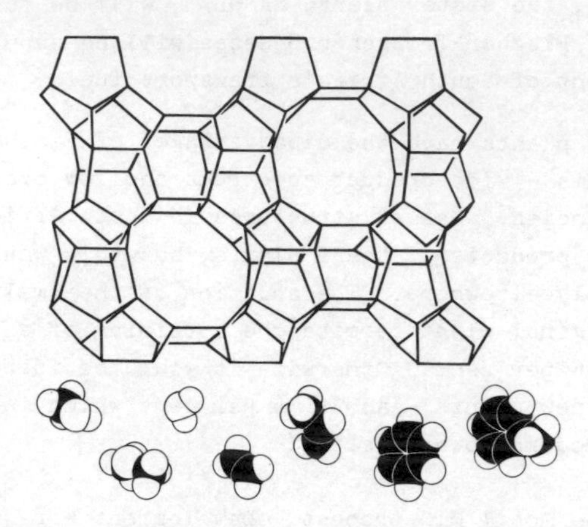

Figure 6 Skeletal diagram of a face of the unit cell of the Mobil zeolite catalyst

Subsequent development work has shown that it is not necessary to isolate methanol or dimethyl ether as intermediate products; methanol synthesis producing raw methanol can be integrated with the Mobil MTG (methanol-to-gasoline) process. The overall efficiency of the coal-to-gasoline only process is about 55 percent. A 100 barrels per day pilot plant, linked to coal gasification, is to be built at Wesseling in West Germany while a commercial 13,000 barrels per day MTG plant using natural gas is to be built in New Zealand.

<u>Direct hydrogenation processes</u>. Although direct hydrogenation processes are potentially more efficient than synthesis processes, the early versions of these processes had several disadvantages including, for example, requirements for very clean (low ash) coals and operating pressures up to 700 bar. Improved processes which reduce the severity of the operating conditions, some through use of catalysts largely developed by the oil industry, are under development, and advances in analytical techniques are being applied with considerable benefit to understanding of the chemistry of the processes.

The most significant work is being carried out in the United States, Germany and Britain with the American programme involving by far the greatest investment. However, there are clear distinctions between the objectives and the approaches adopted.

In general terms, the American processes are designed to produce low-sulphur fuel oils, with naphtha as a co-product. The new German processes yield wide-spectrum synthetic crude oils. Considerable downstream refining of the American and German products would be necessary to convert them to premium fuels. In contrast, the process being developed by the NCB is designed to yield lighter products; principally naphtha and middle distillate.

All direct liquefaction processes involve a hydrogenation stage and the removal of mineral matter. The major differences between processes arise from the order and manner whereby these are accomplished. In the NCB process, removal of mineral matter precedes hydrogenation. This permits use of a highly specific catalyst, which is then less prone to deactivation, for the hydrogenation stage.

The NCB process involves three main unit operations: solvent extraction, hydrocracking and refining. Solvent extraction to obtain 'pure' coal extracts is accomplished using either a liquid or a highly compressed gas as the solvent. Hydrocracking converts these extracts to naphtha and middle distillate for refining to products which may be either petrol and diesel fuel or petrochemical feedstocks.

The Liquid Solvent Extraction variant (Figure 7) can be applied to a very wide range of coal types. Extraction, achieved using a heavy coal-derived solvent, is carried out at pressures below 10 bar. The solvent contains components capable of transferring hydrogen to the coal during the pyrolysis/solution stage and permits depths of extraction up to 90 per cent of the coal substance. The extract solution is filtered free of residues and then passes to the hydrocracker. The solvent's hydrogen donor capability is regenerated in the hydrocracking operation before it is returned to the extraction stage. The thermal efficiency of the process is 70 per cent and the estimated capital cost of a full scale plant is comparable to similar estimates relating to processes being developed in the USA.

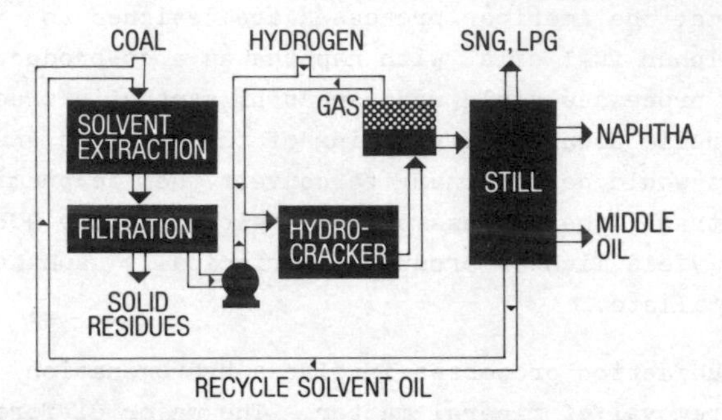

Figure 7 The NCB Liquid Solvent Extraction process

The alternative approach employs a light liquid solvent at a temperature above its critical temperature so that it behaves as a gas. The solvent power of the gas is then a function of its density[12] which is dependent on the pressure, as shown for propane in Figure 8.

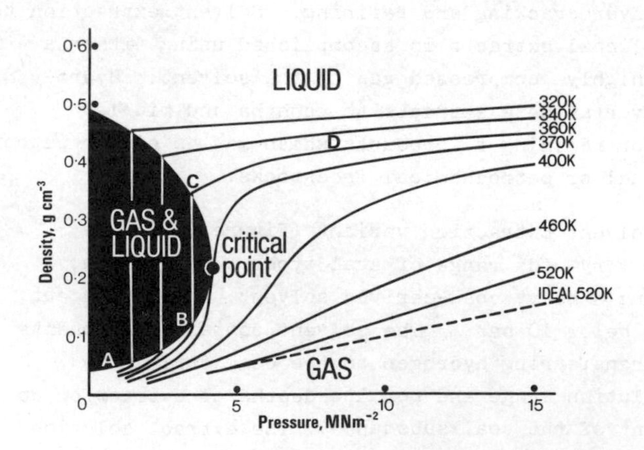

Figure 8 The density of propane as a function of pressure at different temperatures

There are several manifestations of this solvent power of gases in nature. For instance, many natural gas reservoirs have a high level of normally liquid materials in solution in the gas, and when the pressure is released, the liquids separate as so-called 'condensate'.

It is a happy chance that the critical temperatures of many potential solvents for coal lie close to the range of temperatures where thermal breakdown of coals occurs. Careful choice of solvent ensures that it will be in the supercritical state under the extraction conditions. The depth of extraction can be varied considerably, but is generally much less than in liquid solvent extraction. This has the effect of separating the hydrogen-rich fraction of the coal substance as the extract. Removal of residues from the solution is effected by a combination of gas/solid separation operations, including filtration. The extract is readily recovered from the gaseous solvent by decompression and the solvent is recycled. The extract is then combined with a heavy fraction recycled from the hydrocracker before entering the hydrocracker. The proportion of naphtha in the liquid products is higher than it is for the liquid solvent extraction process.

The depth of extraction of coals with supercritical gases is readily adjusted within the range of less than 10 per cent to over 50 per cent. In effect, a fractional separation by extraction of successively higher molecular weight species can be achieved. The distributions of different species and molecular weight fractions at different depths of extraction allow us to characterise components and our understanding of the different structures characterising coals of different ranks is thereby enhanced[13].

The structures originally present in coals are partly preserved in the components of the coal extracts and hydrocracking of these extracts leads to naphthas and middle distillates which have associations with the original structures also. The naphthas are rich in naphthenes. After vapour phase hydrofining to reduce nitrogen and sulphur components to acceptable levels, they can be reformed to high octane gasoline. The mid-distillates are mixtures of bi- and tricyclic hydroaromatics, naphthenes and aromatics. Like the naphthas, they are low in heteroatom content. Further saturation by hydrogenation produces a satisfactory diesel fuel which, because it has a high naphthenes content compared with conventional diesel fuel, has a very low freezing point making it particularly suitable for use in winter or in cold climate zones.

Chemical feedstocks from coal. The routes to petrochemical feedstocks from coal are generally similar to, or extensions of, those leading to premium fuels. There is one exception; the route to vinyl chloride and vinyl acetate through acetylene. The carbide route to acetylene was abandoned elsewhere several years ago on economic grounds but it has been revived in South Africa where the availability of ethylene feedstocks has been constrained. A carbide plant, which consumes over 100,000 tonnes anthracite per year, has been built as part of the petrochemical complex at Sasolburg. Acetylene from this plant is converted to vinyl chloride and about 300 tonnes pvc is produced daily.

Chemicals via gasification and synthesis. Gasification and synthesis routes to ammonia and to methanol have been mentioned already. The Mobil-MTG process yields hydrocarbons, including the lower aromatics, which are equally valuable as chemicals or as components of gasoline. It is possible that the process could be modified to give high yields of olefins. Conventional Fischer-Tropsch processes yield ethylene as well as oxygenated chemicals, i.e. alcohols and acids. Greater selectivity in Fischer-Tropsch synthesis for the production of important chemical intermediates has been demonstrated[14].

Chemicals via hydrogenation and cracking. The naphthas produced by direct hydrogenation processes have a high proportion of naphthenic and aromatic hydrocarbons which can be converted to the most useful aromatic hydrocarbons by straightforward reforming processes. The highly naphthenic middle distillate products give high yields of the lower olefins, together with aromatics, when subjected to conventional cracking processes used in the production of ethylene[15].

Conclusion - Coal in a New Role

The relative reserves of coal, oil and natural gas are such that coal must play a progressively greater role in satisfying the world's energy requirements. In order to fulfil this role, it will be necessary to produce coal more efficiently and use it more effectively, in both cases on a greatly expanded scale. Full advantage can be taken of coal's flexibility as a fuel and of its potential for conversion to liquid and gaseous fuels and chemical feedstocks.

Much work has already been done in increasing current coal usage and in developing the potential for further substantial expansion, through direct substitution for oil and development of improved methods of combustion and efficient methods of coal conversion. A number of the required technologies are already available; of the remainder, most can be proved commercially before the end of the century.

References

1. C.L. Wilson (ed.), "Coal - Bridge to the Future", Report of the World Coal Study, Ballinger, Cambridge, Mass., 1980.

2. J.A. Shearer, G.W. Smith, K.M. Myles and I. Johnson, J. Air Pollution Control Assoc., 1980, 30, Part 6, 684.

3. L.J. Minnick, US/DOE/ET/10415 - T1 Quarterly Technical Progress Report, March-May, 1980.

4. J.W. Nebgen, J.G. Edwards and D. Conway, US/DOE/FHWA-RD 77-136 Final Report, September, 1977.

5. G.F. Morrison, IEA Coal Research Report No. ICTIS/TR11, 1980.

6. M.Y. Lim, IEA Coal Research Report No. ICTIS/TR05, 1979.

7. A.B. Hart, R.S.C. Symposium 'Energy and Chemistry', Paper I.11, 1981.

8. Development of oil and gas resources of the United Kingdom. Dept. of Energy, 1980.

9. J.C. Hoogendoorn, Proc. 9th. Synthetic Pipeline Gas Symposium. Chicago, 1977, p.303 (AGA).

10. K.R. Tart and T.W.A. Rampling, Paper to Coal Chem 2000 (Sheffield, 1980), Inst. Chem.Eng. Symp. Series No.62.

11. R.I. Berry, Chem. Eng., 1980, 87, (8), 86.

12. P.F.M. Paul and W.S. Wise, Principles of Gas Extraction (London, Mills and Boon, 1971).

13. T.G. Martin and D.F. Williams, 'The Chemical Nature of SGE from Low Rank Coals', Royal Society Discussion Meeting, May 1980.

14. Y.S. Tsai, F.V. Hanson, A.G. Oblad and C.H. Yang. "The Hydrogenation of Carbon Monoxide Over Unsupported Iron-Manganese Catalysts for the Production of Low Molecular Weight Olefins", American Chemical Society Houston Meeting 1980.

15. G.O. Davies, 'Making Olefins and Aromatics from Coal Liquids', Coal Chem 2000 (Sheffield, 1980). Inst. Chem. Eng. Symp. Series No. 62.

Stable Coal–Oil Dispersions: A Substitute for Liquid Fuels

By A. J. Groszek
BP RESEARCH CENTRE, CHERTSEY ROAD, SUNBURY-ON-THAMES, MIDDLESEX
TW16 7LN, U.K.

Introduction

Gradual replacement of oil by coal is now widely anticipated by most energy experts. This process has already begun and is likely to continue over the next 50 years or so. Eventually, most of the oil currently used for steam raising and space heating is expected to be displaced by coal and other carbonaceous solids such as coke, shale, coal extracts and wood.

Direct use of solid fuels encounters many disadvantages. Their transportation and distribution is relatively costly compared with that of liquid fuels. There is also a storage difficulty. Many establishments that used coal in the past, but switched to fuel oil when it became cheap and plentiful, disposed of their storage space for coal and have adapted their fuel utilisation systems to the reception of the oil supplied by pipelines and/or road, rail or sea tankers. This adaptation followed the installation of oil-burning plants, many of which are quite new and capable of operation for at least another 10 to 20 years.

Coal-oil mixtures offer a means of saving in the short term up to 40 per cent of fuel oil used for steam raising, blast furnace injection, furnace operation, etc. Also for equipment that has been converted from coal to oil, the use of coal-oil mixtures can be the first stage of returning back to coal.

BP were involved with coal-oil mixtures in the early 1960s when much work was done on coal-oil slurries as a blast furnace injectant[1]. However, the then prevailing low prices of fuel oil precluded further development of this work.

In the mid-1970s the situation changed and it became evident that in the event of restricted fuel oil supply there was not the infra-structure nor the firing equipment available to enable a rapid and convenient changeover from oil to coal to be made.

Past experience with coal-oil mixtures has shown that there are problems with such fuels, namely the necessity for on-site preparation, insufficient stability and combustion particulate deposits. For coal-oil mixtures to become a viable interim technology whilst long-term conversions back to coal are made, it was felt that this type of coal-containing fuel should be stable at the temperatures likely to be encountered in typical storage distribution and firing systems. Also, because such fuels would have inherent stability, they could be prepared at large central facilities and distributed to customers. This would then overcome the problem of customers having to handle both coal and oil as well as the finished product. It was our aim therefore to produce a stable coal-containing fuel, at an economic price, that could be transported, stored and fired in existing equipment designed primarily for fuel oil.

This paper describes the production of very stable coal/fuel oil dispersions and discusses their stability and rheological characteristics in comparison to coal-oil mixtures incorporating an additive. Several trials are described in which stable dispersions of coal in heavy fuel oil dispersion were prepared, then transported in conventional road tankers and fired in burner test rigs.

Type of Coal Dispersions and General Principles Governing Their Stability

The properties of suspensions of coal in fuel oils have recently been extensively investigated in the USA and Japan. The technology studied in most cases was concerned with the stabilisation of mixtures of powdered coal and fuel oils with additives. Water has also been studied as a component of the mixtures and is considered to play an active role in the stabilisation of coal dispersions. Recent fundamental work shows that different coals display varying degrees of affinity for fuel oils, ie coals are classified as being predominantly oleophilic or hydrophilic[2]. This is closely related to the wettability of coals by fuel oils and the type of interaction that occurs when additives are used to stabilise coal-oil mixtures. All the evidence to date indicates that the mechanism of stabilisation of coal-oil mixtures by additives and water depends on the surface properties of coal and the affinity of the surfaces for oil.

Generally, the surface of each particle of coal in coal powders is like a mosaic of hydrophilic and oleophilic (or hydrophobic) sites. The hydrophilic sites result from the presence of oxygen-containing functional groups as shown recently by Perry and Grint[3] and also mineral inclusions on the surface. Oleophilic sites are more difficult to characterise but are probably due to various organic radicals such as methyl and methylene groups as well as polycyclic aromatic elements within the main structure of coal. Thus, a typical particle of coal may be represented schematically as shown in Figure 1. where individual sites alternate between polar and non-polar sites. This situation can be altered in the presence of surface active agents (SAA) that can be added to the liquid fuel in which the particles are suspended. The agents tend to adsorb on the surfaces of the particles as shown in Figure 2. The polar sites on the coal particles are then converted to oleophilic sites, which increases the affinity of the coal surface for the liquid hydrocarbon in which it is dispersed. With appropriate additives, reasonably stable dispersions may be obtained in this way as recently reported by the Japanese[4] and American[2] workers.

Water plays an important role in the stabilisation of coal-oil mixtures as it combines with the polar sites on the surfaces of coal particles and can link them most probably with the aid of hydrogen bonding. This type of aggregation is shown schematically in Figure 3. Additives used for the stabilisation of coal-oil mixtures are rendered more effective in the presence of water[2] presumably due to stronger and more complete coverage of coal surfaces when both water and surface active compounds are present. Both water and surface active agents can have stabilising or destabilising action on the suspensions of coal in fuel oils depending very much on the relative proportions of hydrophilic and oleophilic sites on

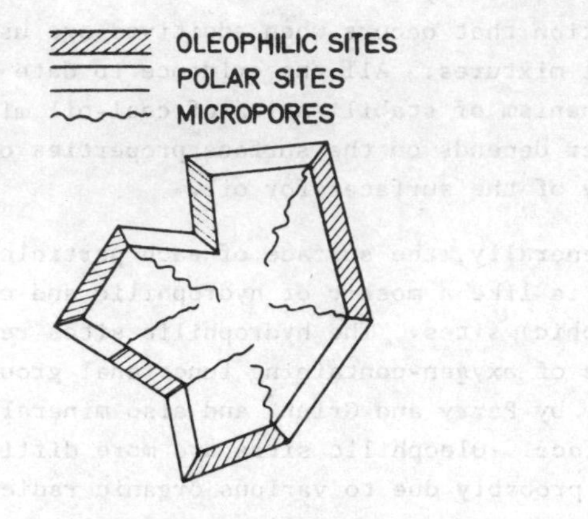

FIGURE 1
A COAL PARTICLE WITH
DIFFERENT SURFACE SITES

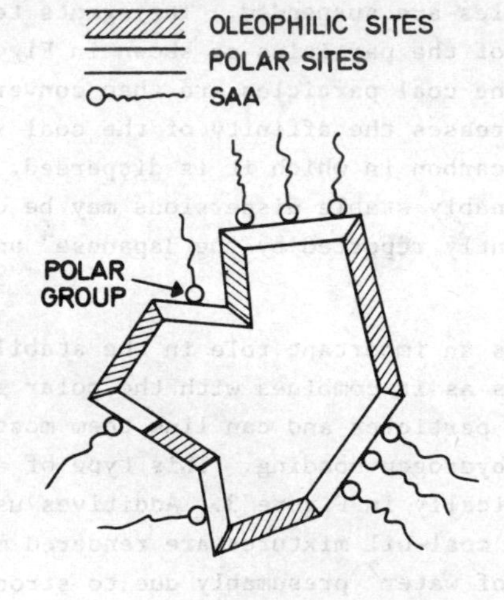

FIGURE 2
ADSORPTION OF SURFACE ACTIVE AGENTS ONTO
POLAR SITES OF A COAL PARTICLE

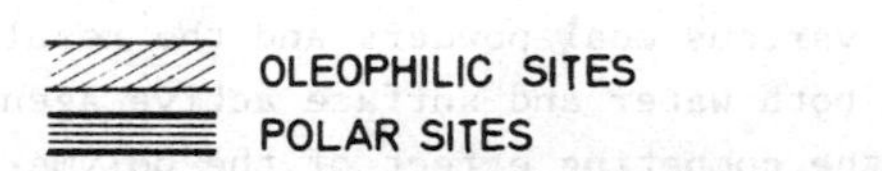

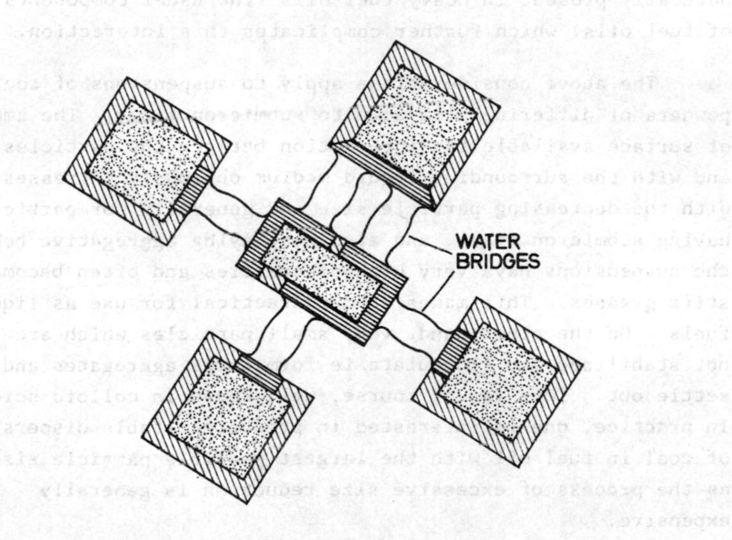

FIGURE 3
BRIDGING OF COAL PARTICLES BY WATER
VIA POLAR SITES

the particles. For very hydrophilic powders water can cause rapid flocculation and precipitation due to complete wetting of the particles which then become surrounded by water, leading to their coalescence and rapid segregation[2]. Botsaris et al. concluded that coal-oil mixtures are exceedingly complex systems mainly because of the variability in the surface character of particles of various coal powders and the resulting differences in the way that both water and surface active agents is complicated by the competing effect of the polymer compounds naturally present in heavy fuel oils (the usual components of fuel oils) which further complicates this interaction.

The above considerations apply to suspensions of coal powders of differing size down to submicron sizes. The amount of surface available for interaction between the particles and with the surrounding liquid medium obviously increases with the decreasing particle size and generally for particles having submicron sizes, and also displaying aggregative behaviour, the suspensions have very high viscosities and often become stiff greases. This makes them impractical for use as liquid fuels. On the other hand, very small particles which are not stabilised may flocculate, ie form large aggregates and settle out. This is, of course, well known in colloid science. In practice, one is interested in producing stable dispersions of coal in fuel oil with the largest possible particle sizes as the process of excessive size reduction is generally expensive.

The knowledge of the surface chemistry of coal should enable one to realise the following main objectives in the production of coal-oil fuels

1. stability at a range of temperatures relevant to practical usage,
2. acceptably low viscosity commensurate with ease of pumping and utilisation in combustion systems,
3. high content of coal.

The last objective opposes the aim stated under point 2 and in practice a compromise has to be reached to obtain an acceptable system.

BP Route to Preparation of Very Stable Coal/Fuel Oil Dispersions

Based on fundamental studies of solid-liquid interactions that have been carried out at the BP Research Centre, a novel method of preparing very stable coal/fuel oil dispersions has been developed which promotes the formation of exceptionally stable dispersions in a wide range of fuel oils. The use of stabilising additives (the route adopted by the majority of workers preparing coal-oil mixtures) is not required. Very stable dispersions have been achieved for several combinations of fuel oils and coals. The universal applicability of the process is considered important for its widespread adoption.

The BP process results in the production of dispersions stable over a range of temperatures up to 100°C for long periods of time and retaining the stability during transportation and handling prior to injection into burning equipment. Tests of these stable dispersions in a variety of rigs and small boilers described below indicate that they burn essentially like liquid fuels and are therefore suitable for their replacement.

Stability of Coal/Oil Dispersions in Comparison With Typical Coal Oil Mixtures

To determine the stability of coal-oil dispersions we have used a sedimentation technique on a tube with a facility for measurement of the solids content of the base sample. This stability tube enabled a sample to be removed from its base after holding at a given temperature for a fixed period of time without disturbing the remainder of the tube contents. The supernatant liquor is then decanted off and the dilution/centrifugation step repeated three more times. The remaining solid plug is then dried and weighed.

The stability was then defined as:-

The coal content of base sample minus initial coal content of coal-oil mixture.

The number is expressed in per cent weight points.

Invariably coal-oil mixtures are claimed to be stable at ambient and temperatures up to a maximum of about 71°C (160°F). The most severe static sedimentation test published to date has been undertaken at 100°C for one hour followed by 70°C for two weeks[5]. To be a viable replacement for fuel oil, coal-oil mixtures must be sufficiently stable to withstand the vagaries of unattended distribution, storage and firing systems. For pumping and satisfactory firing, such mixtures (especially those containing heavy fuel oil) have to be heated to temperatures sometimes as high as 150°C (302°F). When a flame out situation arises these fuels may have to be held at such high temperatures, if it is proposed that coal oil mixtures can be used as a replacement for fuel oil alone, with the minimum of modification. In addition, whilst in storage tanks the coal-oil mixture has to be heated, either by steam coils or electric elements, where surface temperatures in excess of 100°C (212°F) are encountered. The incorporation of additional mixing devices to already existing tankage is not an ideal solution to this problem. It is far better to produce a coal-containing fuel that is stable under the conditions it is likely to encounter. It is felt that coal-oil mixtures stabilised with additives of the types commonly used to date will not be sufficiently stable to withstand such temperatures under static conditions.

Utilising pulverised coal and heavy fuel oil (typical inspection data are shown in Tables 1 and 2 respectively), a sample of coal/fuel oil dispersion (35 per cent weight coal) was prepared for the demonstration trial (see later). This dispersion has been compared with two examples of coal-oil mixtures stabilised with additives.

Stability measurements using the BP stability tube have been undertaken on coal oil mixtures (including the additives) at coal concentrations of 35, 40, 45 and 50 per cent weight in comparison to the BP produced material (35 per cent weight coal). At this stage it is important to point out that stability of a coal-oil mixture or dispersions increases with increasing coal content and increasing oil viscosity. A plot of the variation of stability with coal content for a coal-oil mixture (additive route) is shown in Figure 4. The necessity for high

TABLE 1 INSPECTION DATA ON BITUMINOUS COAL USED IN HAMWORTHY COMBUSTION TEST

As Received Analysis		
Ash content	%wt	5.0
Moisture content	%wt	2.5
Total sulphur content	%wt	1.6
Pyritic sulphur content	%wt	(0.8)
Chlorine content	%wt	0.26
Carbon content (total)	%wt	77.9
Hydrogen content	%wt	4.89
Nitrogen content	%wt	1.70
Calorific value (gross)	MJ/kg	32.36
	Btu/lb	13 910
Volatile matter content (estimated dmmf)	%wt	(33 - 34)
Particle Size Analysis		
+500 μm	%wt	0.1
-500 + 212 μm	%wt	1.6
-212 + 75 μm	%wt	21.0
-75 + 53 μm	%wt	14.2
-53 μm	%wt	63.1
Ash Fusibility Data (Semi-Reducing Atmosphere)		
Deformation temperature	°C	1 120
	°F	2 048
Hemisphere temperature	°C	1 260
	°F	2 300
Flow temperature	°C	1 400
	°F	2 552
Ash Analysis		
SiO_2	%wt	42.1
Al_2O_3	%wt	26.5
CaO	%wt	4.5
MgO	%wt	1.08
Fe_2O_3	%wt	15.4
TiO_2	%wt	1.48
Na_2O	%wt	1.60
K_2O	%wt	1.73
Mn_3O_4	%wt	0.25
P_2O_5	%wt	0.14
SO_3	%wt	2.35

Notes:- Values in parenthesis are estimated values

All analyses performed according to BS 1016, except ash analysis (X-ray fluorescence/emission spectroscopy)

dmmf signifies dry mineral matter free

TABLE 2

INSPECTION DATA ON HEAVY FUEL OIL
(SOURCE: MIDDLE EAST CRUDES)
USED IN HAMWORTHY COMBUSTION TEST

Specific gravity at 15.6°C/15.6°C (60°F/60°F)		0.977
Total sulphur content	%wt	3.38
Kinematic viscosity at 60°C (140°F)	cSt	219
" " " 80°C (176°F)	cSt	72.5
" " " 100°C (212°F)	cSt	37.5
Conradson carbon residue (ASTM D189)	%wt	11.2
Water content	%vol	0.10
Asphaltenes (IP 143)	%wt	3.1
Nitrogen content	%wt	0.27
Ash content at 550°C	%wt	0.114
Vanadium content	ppm wt	55
Sodium content	ppm wt	32
Calorific value (gross)	MJ/kg (Btu/lb)	42.2 (18 200)

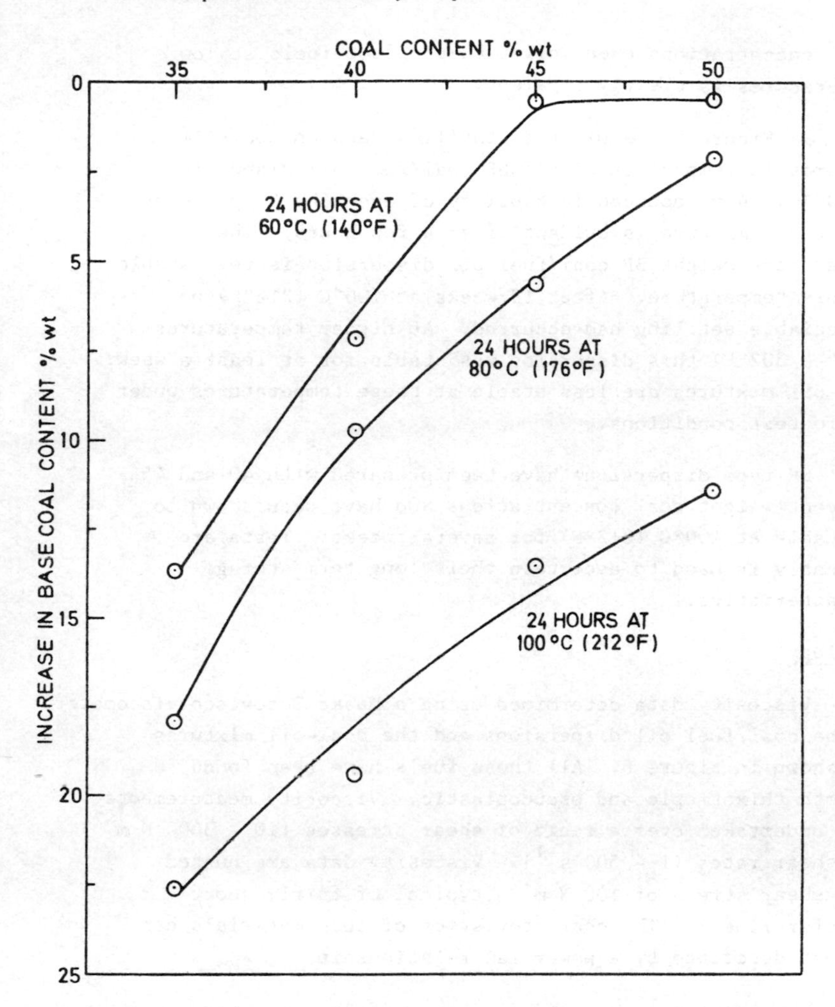

FIGURE 4
THE VARIATION OF STABILITY WITH COAL CONTENT
FOR COAL OIL MIXTURES

coal concentrations even whilst keeping the fuels at low temperatures is clearly apparent.

In Figure 5 are plotted stability data on coal-oil mixtures in comparison with a BP coal/fuel oil dispersion at 100°C. A pronounced instability of even the 50 per cent weight oil mixture is evident after a few hours. The 35 per cent weight BP coal/fuel oil dispersion is very stable at this temperature. After 13 weeks at 100°C (212°F) no appreciable settling had occurred. At higher temperatures (150° - 302°F) this dispersion was stable for at least a week. Coal-oil mixtures are less stable at these temperatures under static test conditions.

BP type dispersions have been prepared with 40 and 45 per cent weight coal concentrations and have been shown to be stable at 100°C (212°F) for several weeks. Tests are currently in hand to ascertain their long term storage characteristics.

Rheology

Viscosity data determined using a Haake Rotovisco viscometer on the coal/fuel oil dispersions and the coal-oil mixtures are shown in Figure 6. All these fuels have been found to be both thixotropic and pseudoplastic. Viscosity measurements were undertaken over a range of shear stresses (10 - 3000 N m^{-2}) and shear rates (1 - 500 s^{-1}). Viscosity data are quoted at a shear stress of 100 N m^{-2} (typical of fairly short transfer lines). The characteristics of such materials can be best described by a power law relationship

$$\sigma = K\dot{\gamma}^{N} \text{ or } \eta_{app} = K\dot{\gamma}^{N-1}$$

σ shear stress
$\dot{\gamma}$ shear rate
η apparent viscosity
N power law index

For a Newtonian fluid N equals 1, for a shear thickening material $N > 1$ whilst for a shear thinning fluid $N < 1$.

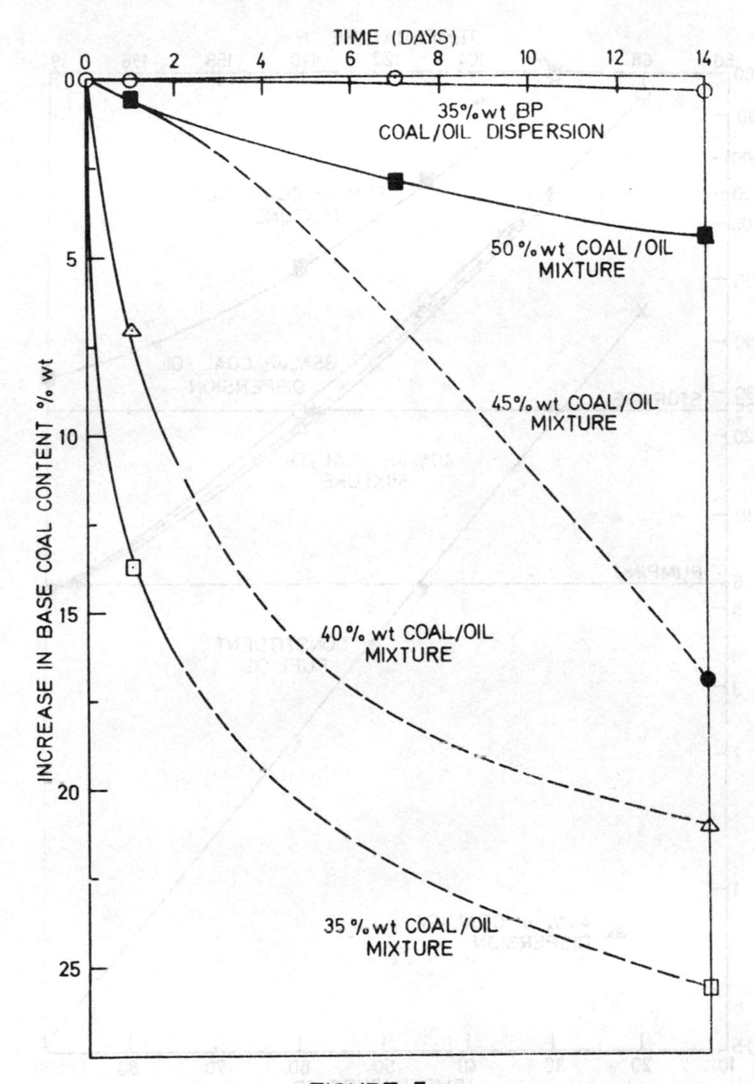

FIGURE 5
THE VARIATION OF STABILITY WITH TIME AT 60°C (140°F)

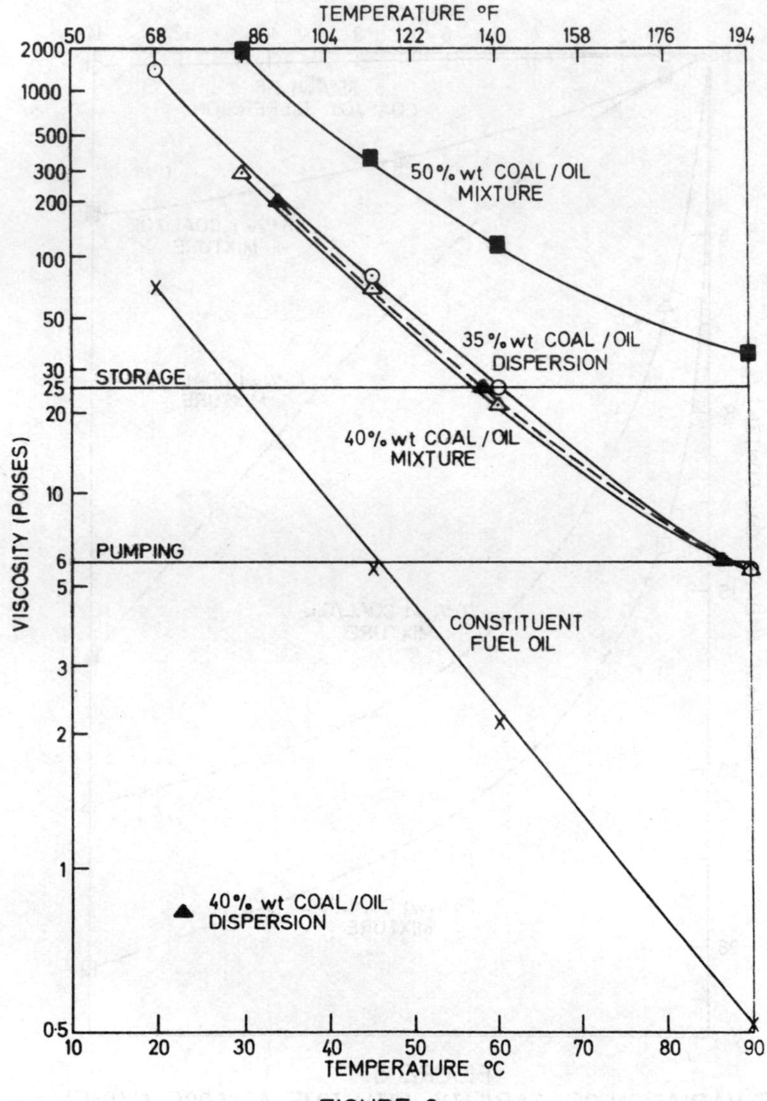

FIGURE 6
VISCOSITY vs TEMPERATURE RELATIONSHIPS

The constant K is the value of the stress at a shear rate of 1 s^{-1}; when N equals 1, K becomes the Newtonian viscosity. Power law parameters and apparent viscosities at 100 N m^{-2} shear stress on the materials evaluated at 45°, 60° and 90°C (113°, 140° and 194°F) are quoted in Table 3. The tabulated values of K and N can be used with the above equations to calculate apparent viscosities for any shear rate or shear stress. Although a single viscosity figure will not fully characterise a non-Newtonian fluid there are times, for example, when making comparisons, where such a figure is useful.

The extent of increase in viscosity of coal containing fuels over that of the constituent fuel oil can be seem from the data in Table 4, where

$$\text{relative viscosity} = \frac{\text{viscosity of coal oil fuel}}{\text{viscosity of fuel oil}}$$

It can be seen that for the 50 per cent weight coal oil mixture a very high relative viscosity was obtained.

On Figure 6 are shown the maximum viscosities recommended for storing (25 poise) and pumping (6 poise) heavy oils (see IP 230 for details). The recommended minimum temperatures based on these criteria are quoted in Table 5. It is evident that 50 per cent weight coal in oil mixtures in heavy fuel oil (370 cSt at 50°C - 3500 seconds Redwood 1 at 100°F) are very viscous and required to be heated to temperatures far higher than their stabilities would allow. However, to reduce coal content also leads to fuels of unsatisfactory stability with the exception of the BP produced materials (Figure 5). Included in Figure 6 is the viscosity temperature plot for a BP dispersion containing 40 per cent weight coal in a 300 cSt at 50°C (2800 seconds Redwood 1 at 100°F) fuel oil.

Scope of the BP Process

Stable coal/fuel oil dispersions (ie 24 hours at 100°C) have been prepared using a variety of fuel oils ranging from 30 cSt at 50°C (200 seconds Redwood 1 at 100°F) to 600 cSt at 50°C (6000 seconds Redwood 1 at 100°F). The coals used

TABLE 3

POWER LAW PARAMETERS AND APPARENT VISCOSITIES AT 100 N m^{-2} SHEAR STRESS FOR THE COAL OIL FUELS

Sample	45°C (113°F)			60°C (140°F)			90°C (194°F)		
	K	N	η_{Pas}	K	N	η_{Pas}	K	N	η_{Pas}
Constituent fuel oil	0.583	0.99	0.556	0.217	1.00	0.217	0.052	1.00	0.052
Additive coal oil mixture 40% wt coal	8.28	0.91	6.51	2.30	0.96	1.93	0.78	0.92	0.517
Additive coal oil mixture 50% wt coal	44.9	0.83	38.1	15.6	0.87	11.67	4.97	0.89	3.36
BP coal/fuel oil dispersion 35% wt coal	8.73	0.94	7.47	2.79	0.96	2.40	0.62	0.97	0.518

Notes:- K - apparent viscosity in Pas at a shear rate of 1 s^{-1}

N - power law index

η - apparent viscosity in Pas at shear stress of 100 N m^{-2}

1 Pas = 10 Poise

TABLE 4

VISCOSITIES OF THE COAL-OIL FUELS RELATIVE TO THE VISCOSITY OF THE CONSTITUENT FUEL OIL

Temperature	Relative Viscosity			
	Additive Coal-Oil Mixture		BP Coal/Fuel Oil Dispersion	
°C (°F)	40% wt Coal	50% wt Coal	35% wt Coal	40% wt Coal
45 (113)	11.7	68.5	13.4	14.1
60 (140)	8.9	53.8	11.1	12.9
90 (194)	9.9	64.6	10.0	11.6

TABLE 5

RECOMMENDED MINIMUM TEMPERATURES FOR STORING AND HANDLING (IP 230)

Sample	Temperature °C (°F)	
	Storage (25 Poise)	Handling (6 Poise)
Constituent fuel oil	28 (82)	44 (111)
BP coal/fuel oil dispersion (35% wt coal)	58 (136)	86 (187)
BP coal/fuel oil dispersion (40% wt coal)	58 (136)	86 (187)
Additive coal-oil mixture (40% wt coal)	57 (135)	85 (185)
Additive coal-oil mixture (50% wt coal)	100 (212)	>120 (>248)

to prepare such stable mixtures have included bituminous coals, sub-bituminous coals and a lignite. The ash contents of these materials varied between 2.7 and 26.4 per cent weight, sulphur contents between 0.3 and 4.1 per cent weight, water contents between 0.8 and 17.5 per cent weight. Generally, our method is suitable for the preparation of stable dispersions of all types of carbonaceous solids, as summarised in Table 6.

Coal/Fuel Oil Dispersion Demonstration Trial

A test to demonstrate the feasibility of producing a stable coal/fuel oil dispersion at a central location, then transporting, storing and firing this material in a system designed primarily for heavy fuel oil was carried out in 1978 when 50 tons of a 35 per cent coal dispersion were prepared using a heavy fuel oil as the liquid component. This material was then transported in conventional fuel oil road tankers to storage tanks located 50 miles away from the preparation site. After storage for six weeks (nominally at 65°C - 149°F), the coal/fuel oil dispersion was then transported, again by road tanker, 140 miles to Poole, Dorset, and fired in a burner test rig. The timetable and location of each stage of this demonstration trial are as follows:-

July 1978:	Preparation	- BP Chemicals Limited Carshalton, London
August 1978:	Handling and Storage	- BP Oil Kent Refinery Ltd Isle of Grain, Kent
September 1978:	Firing	- Hamworthy Engineering Ltd Poole, Dorset

(see Figure 7)

Preparation

Stable coal in oil dispersions can be prepared as shown diagrammatically in Figure 8. The pulverised coal is fed by screw feeder to the top of the 7000 litre steel vessel already containing heavy fuel oil at 64° - 70°C (147° - 158°F) fed from the road tanker. Coal and heavy fuel oil are the only two components used in the preparation route. The coal oil mixture is continually agitated using a single speed, twin paddle stirrer, and is then fed to the milling equipment

TABLE 6

LIQUID COMPONENTS	60 - 70% wt
FUEL OILS	
CRUDE OILS	
COAL PYROLYSIS LIQUIDS	
SYNCRUDES	
COMBUSTIBLE SOLID COMPONENTS	**40 - 30% wt**
COALS	
BROWN	
SUB-BITUMINOUS	
BITUMINOUS	
COKES	
FLEXICOKE	
DELAYED COKE	
PYROLYSIS RESIDUES	
PITCHES AND ASPHALTS	

TABLE 7 COAL/FUEL OIL DISPERSION

Coal content	%wt	35.4
Ash content	%wt	1.78
Nitrogen content	%wt	0.78
Sulphur content	%wt	2.75
Ash Fusibility (Semi-Reducing Atmosphere) BS 1016		
Deformation temperature	°C (°F)	1070 (1958)
Hemisphere temperature	°C (°F)	1240 (2264)
Flow temperature	°C (°F)	1420 (2588)
Approximate relative density at 20°C (68°F)		1.1
" " " " 120°C (248°F)		1.0
Apparent Viscosities at Shear Stress 100 N m^{-2} at		
20°C (68°F)	Poise (cSt)	1330 (120 910)
45°C (113°F)	Poise (cSt)	74.6 (6 940)
60°C (140°F)	Poise (cSt)	22.7 (2 140)
90°C (194°F)	Poise (cSt)	5.05 (490)
120°C (248°F)	Poise (cSt)	2.00 (200)

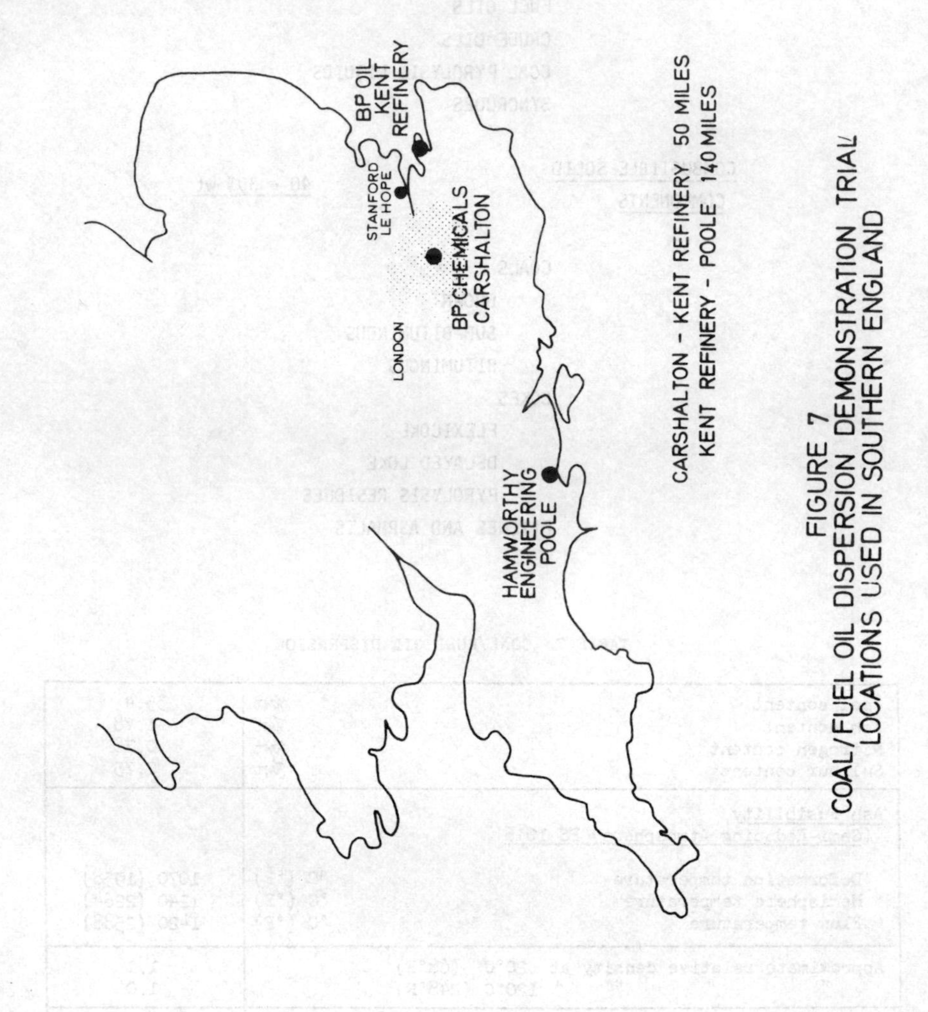

FIGURE 7
COAL/FUEL OIL DISPERSION DEMONSTRATION TRIAL LOCATIONS USED IN SOUTHERN ENGLAND

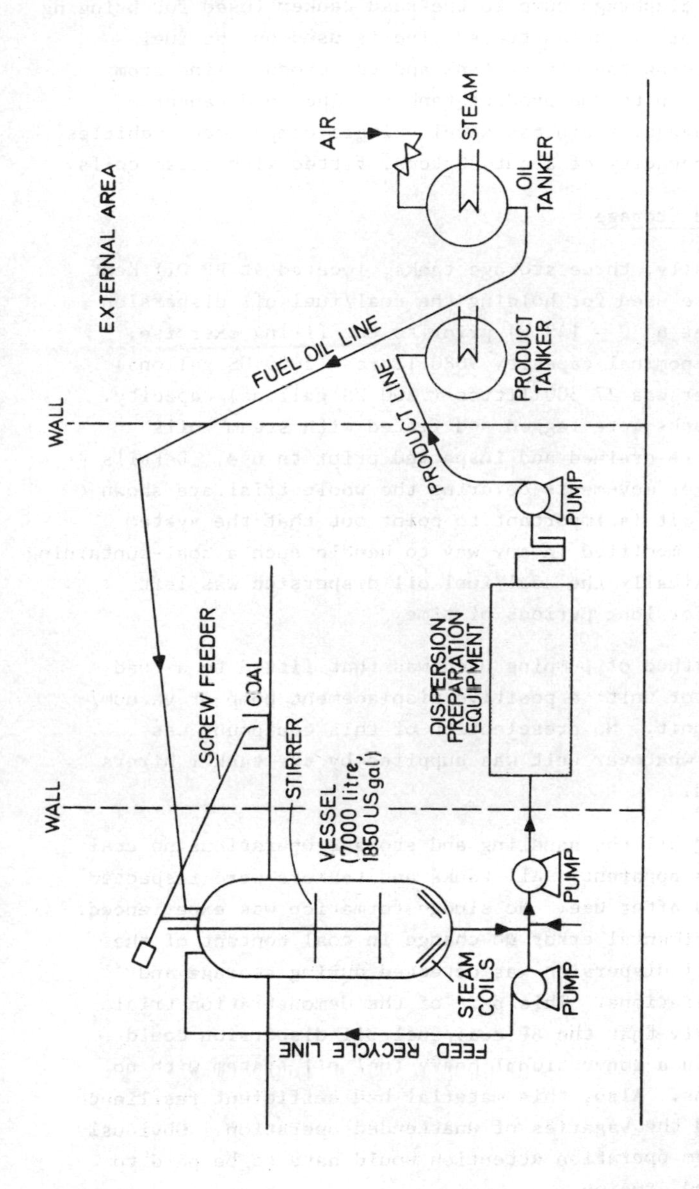

FIGURE 8
ELEVATION DIAGRAM OF CARSHALTON RIG FOR PREPARING COAL/FUEL OIL DISPERSION

using a diaphragm pump. The stable dispersion is then pumped via another diaphragm pump to the road tanker (used for bringing in the fuel oil). Steam traced line is used on the fuel oil line feeding the mixing tank and the product line from the second pump to the product tanker. The road tankers used were lagged, stainless steel, single compartment vehicles of nominal capacity of about 20 tons, fitted with steam coils.

Handling and Storage

Initially, three storage tanks, located at BP Oil Kent Refinery were used for holding the coal/fuel oil dispersion (nominally at 65°C - 149°F) prior to the firing exercise. Two were of nominal capacity 9080 litres (2400 US gallons) and the other was 27 300 litres (7200 US gallons) capacity. All three tanks were lagged and fitted with steam coils. The tanks were drained and inspected prior to use. Details on road tanker movements covering the whole trial are shown in Table 8. It is important to point out that the system used was not modified in any way to handle such a coal-containing fuel and basically the coal/fuel oil dispersion was left unattended for long periods of time.

The method of pumping used was that fitted to a road tanker tractor unit: a positive displacement pump or vacuum/ compressor unit. No preselection of this equipment was undertaken: whatever unit was supplied by the tanker hirers was utilised.

During all the handling and storage operations no coal settling was apparent. All tanks and tankers were inspected prior to and after use. No sludge formation was experienced. Within experimental error no change in coal content of the coal/fuel oil dispersion was detected during storage and handling operations. This part of the demonstration trial showed clearly that the BP coal/fuel oil dispersion could be handled in a conventional heavy fuel oil system with no modifications. Also, this material had sufficient resilience to withstand the vagaries of unattended operation. Obviously for long-term operation attention would have to be paid to pump wear and erosion.

TABLE 8

HANDLING AND STORAGE OF 46 000 LITRES (12 150 US GAL) OF COAL/FUEL OIL DISPERSION JULY/SEPTEMBER 1978

Date	Tanker No	Journey	Quantity Discharged litre (USgal)	Time for Discharge min	Average Discharge Rate litre/h (USgal/h)	Approximate Temperature of Dispersion °C (°F)	Method of Pumping	Approximate Pipework Dimensions	
								Diameter mm (in)	Length m (ft)
24th July	1	Carshalton - Kent Refinery	18 150 (4798)	117	9 340 (2468)	60 (140)	PD pump	76 (3)	21.3 (70)
30th July	2	Carshalton - Stanford le Hope	←			*			→
3rd August	3	Carshalton - Stanford le Hope	←			*			→
18th August	2 3	Stanford le Hope - Kent Refinery	18 600 (4914) 9 530 (2518)	85 33.5	13 129 (3469) 17 068 (4509)	60 (140) 58 (136)	PD pump PD pump	76 (3) 76 (3)	24.4 (80) 24.4 (80)
5th September	4	Kent - Poole	17 600 (4650)	480	2 200 (581)	60 (140)	Vacuum	76 (3)	21.3 (70)
11th September	5 6	Kent - Poole	17 365 (4588) 10 080 (2663)	137 81	7 606 (2010) 7 467 (1973)	103 (217) 103 (217)	PD pump PD pump	76 (3) 76 (3)	24.4 (80) 24.4 (80)
12th September	5	Contents transferred to Tanker No 4	17 320 (4576)	50	20 785 (5491)	79 (174)	PD pump	63.5 (2.5)	18.3 (60)
13th September	6	Contents transferred to Tanker No 4	10 080 (2663)	110	5 499 (1453)	50 (122)	Compressed air	63.5 (2.5)	18.3 (60)

Notes:- *Maintained on steam at depot in tanker trailer

PD - positive displacement

Subsequently, a larger 500 ton batch of stable coal oil containing 40 per cent of coal in fuel oil was prepared and its pumpability and rheology examined in some detail. There was good agreement between the pressure drops measured during the pumping trials and the values calculated using viscosity data determined in the laboratory. The pressure drops were measured over a 5 m test section of nominal 3 inch standard unlagged flexible tanker hose. The dispersion was pumped from one tanker (4000 gallons capacity) to another through the test section. The trial showed that the 40 per cent dispersion could be handled at 50°C and complete transfers achieved from one tanker to another, albeit at a much slower rate than that obtained for fuel oil alone. However, at 65 - 70°C pumping rates were comparable to those normally obtained for the heavy fuel oils Otherwise, if pumping had to be carried out at temperatures around 50°C, pipe diameters larger than 3 inches would have to be used to achieve acceptable pumping rates.

Storage of the 500 ton batch of the 40 per cent coal dispersion has now continued for some nine months at temperatures ranging from 40° to 70°C with no measurable signs of settling.

Combustion in a Burner Test Rig

In order to assess the combustion characteristics of the coal/fuel oil dispersion, approximately 56 tons were fired (nominally 100 hours continuous operation) in a marine burner test rig at the premises of a burner manufacturer (Hamworthy Engineering Limited). For comparative purposes a similar quantity of the constituent fuel oil was fired also. A marine boiler was chosen since stable coal/fuel oil dispersions show great potential as a possible marine fuel. It is thought unlikely that coal-fired ships of the type seen before the widespread use of fuel oil would again be tolerated. A modern method of handling coal, clean and not manpower intensive, will be a major requirement of operators.

The burner chosen for the firing exercise was a Hamworthy/Wallsend HX 360 marine register with an external mix steam assisted pressure jet atomiser. This type of burner is in service with BP Tanker Company and other shipping companies.

The test rig consists of a water cooled horizontal steel chamber with natural convection circulation.

The schematic arrangement of the fuel system is also shown in Figure 9. It can be seen that this system is typical of that used for heavy fuel oil with the additional facility of a road tanker for storing the coal/fuel oil dispersion.

Operating data obtained whilst burning the coal/fuel oil dispersion and its constituent fuel oil for comparison, both fired at a nominal flow rate of 455 kg/h (1000 lb/h) for 100 hours, are shown in Table 9. The main visual difference between the flames was that with the dispersion the centre core of the flame was surrounded with a concentric ring of flames when viewed from the rear. Both flames were of similar dimensions, 2.2 metres (7.2 feet) long with a maximum diameter of 1.3 metre (4.3 feet). A 'golden rain' (fine sparklers) was visible when firing the coal containing fuel.

Particulate and gaseous emissions obtained during the combustion trial are quoted in Tables 10 and 11 respectively. As to be expected, particulates and nitrogen oxides were higher for the dispersion, whilst sulphur dioxide levels were reduced (the coal contained less sulphur than the oil). Approximately, 10 per cent weight of the ash contained in the coal/fuel oil dispersion was deposited in the test chamber. The deposits were dry, red in colour and had the texture of talcum powder, and were removed after the test by means of a vacuum cleaner. No slagging of the ash deposits was evident. The ash retained in the combustion chamber reduced the heat to water transfer rate which caused the combustion gas temperature at the superheater inlet to rise by 200°C (392°F). However, it must be pointed out that the gas velocities in the test rig combustion chamber were a factor of four lower than in a marine boiler. Therefore, the deposition results obtained from the rig are unlikely to be representative of those obtained in practice. Although superheater tube deposition was experienced on the rig, this was not considered a problem on convective superheaters, as in practice steam soot blowers should easily remove the type of ash deposits obtained when firing the BP coal/fuel oil dispersion.

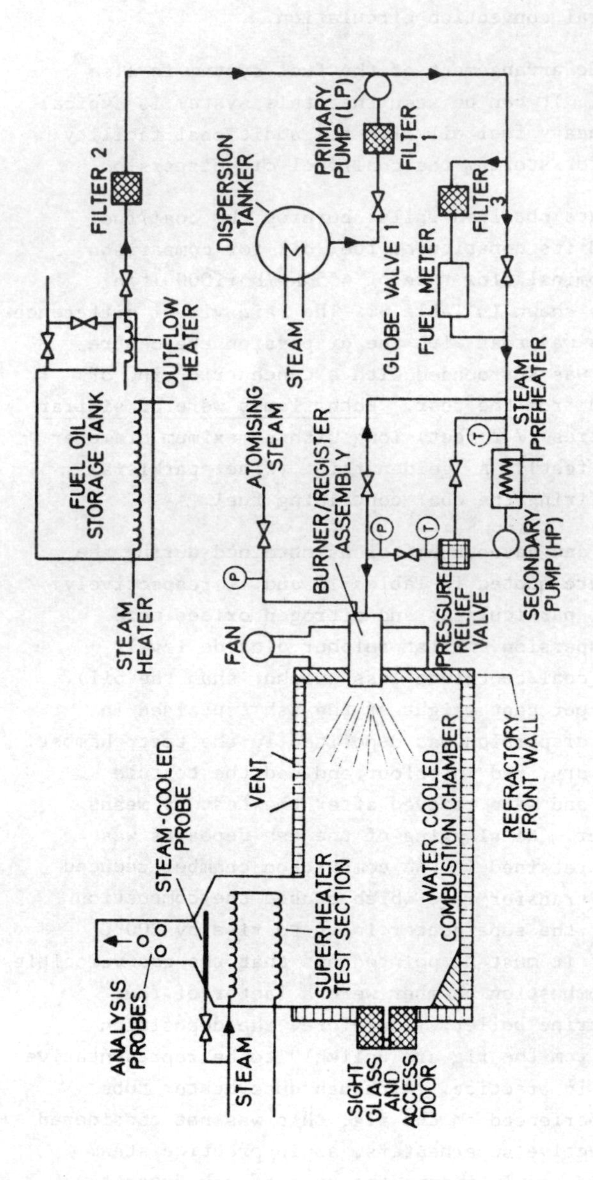

FIGURE 9
SCHEMATIC ARRANGEMENT OF HAMWORTHY TEST RIG FOR COMBUSTION TRIAL OF THE COAL/FUEL OIL DISPERSION

TABLE 9

BURNER TEST RIG OPERATION DATA

	Flow Rate kg/h	Fuel Temperature at Burner °C (°F)	Fuel Pressure at Burner bar (ga)	Fuel Pressure at Pump Outlet bar (ga)	Fuel Pressure at Line Heater Outlet bar (ga)	Fuel Temperature at Heater Inlet °C (°F)	Atomising Steam Pressure at Burner bar (ga)	Uptake Flue Gas Temperature °C (°F)	Windbox Pressure mbar	Total Firing Time h	Total Quantity of Fuel Burnt litre (USgal)
Dispersion											
Minimum	472	142 (288)	15.86	14.48	2.90	60 (140)	3.03	700 (1292)	8.47		
Maximum	500	148 (298)	17.93	22.06	4.48	86 (187)	3.03	900 (1652)	9.96	95.2	43 060 (11 380)
Mean	483	144.4 (293)	17.00	18.00	3.32	78.6 (173)	3.03	801 (1474)	8.97		
Fuel Oil											
Minimum	388	130 (266)	23.44	24.82	3.10	82 (180)	2.14	650 (1202)	8.97		
Maximum	444	136 (277)	34.48	30.89	3.45	114 (237)	2.48	725 (1337)	8.99	88.7	39 154 (10 344)
Mean	419	132.8 (270)	25.51	27.44	3.30	99.4 (211)	2.24	694.2 (1281)	8.97		

Note:- 1 bar ≡ 14.5 lb/in^2

TABLE 10

BURNER TEST RIG PARTICULATE EMISSIONS AND SMOKE NO

		Coal/Oil Dispersion			Constituent Fuel Oil		
Date		12.9.78	14.9.78	Mean	26.9.78	28.9.78	Mean
Particulates %wt on fuel		1.47 1.31 1.39 1.58 1.66	1.40 1.47 1.29 1.51 1.40	1.45	0.07 0.10 0.08 0.10	0.07 0.08 0.11 0.10	0.09
Smoke No		6	7	6.5	4	4	4
Excess oxygen	%vol	4.1	5.2	4.6	4.5	5.6	5.0

TABLE 11

BURNER TEST RIG GASEOUS EMISSIONS

		Coal/Oil Dispersion	Constituent Fuel Oil
O_2	%vol	3.9	5.2
CO_2	%vol	13.3	12.0
CO	ppm vol	40	75
NO_x	ppm vol	325	210
SO_2	ppm vol	1534	1696
SO_3	ppm vol	39	37
Conversion of SO_2 - SO_3	%	2.5	2.2

After firing all the coal/fuel oil dispersion the fuel system was dismantled and inspected. No coal deposits were identified at any point. The stability of the BP dispersion was well demonstrated when the burner nozzle was being cleaned. Dispersion held at 145°C (293°F) gave rise to no coal settling. No problems of coal deposition in the steam preheater were encountered even though the steam temperature was 188°C (370°F).

During the assembly of equipment for this trial, no attempt had been made to select items for use with coal; therefore it was not surprising that significant wear was encountered in the two gear pumps, pressure relief valve and burner nozzle. Suitable equipment for handling coal oil fuels is already available and its use will eliminate the wear and erosion problems.

Comparison of a Coal Oil Dispersion and Coal Oil Mixture in a Test Furnace Carried Out by the International Flame Research Foundation

This test was carried out recently to compare directly the combustion characteristics of a coal-oil dispersion with a coal-oil mixture (COM). The dispersion was prepared according to the BP method and the COM by simple mixing of powdered coal (of the type used in coal burning power stations) and heavy fuel oil. Coal-oil mixtures have been extensively tested in the USA with the support of the Department of Energy, using modified burners and equipment which enabled the operators to burn the mixture immediately after mixing.

In the IFRF test the fuels were fired in a furnace fitted with cooling tubes which created a low intensity water tube boiler environment. The burner used was fitted with two different atomiser types (a Y-Jet and an internal mix type). Data obtained on optimised flames are summarised in Table 12.

No difficulty was experienced in optimising the burner parameters to burn the stable BP dispersion under conditions similar to those for heavy fuel oil, using the standard burner gun, fitted with both Y-Jet and internal mix atomiser.

TABLE 12

DATA OBTAINED ON OPTIMISED FLAMES

	HFO	COD	HFO	COD	COM
Flame No	0-3	D-12	0-2	D-2'	M-6
Atomizer Type	Y-Jet	Y-Jet	1M	1M	1M (modified)
Spray Angle°	60	60	70	70	70
Excess Air % vol	15	15	20	20	30 (more critical)
Swirl No	0.4	0.5	0.5	0.5	0.5
Atomizing Steam wt/wt on fuel	0.1	0.14	0.12	0.18	0.21
Fuel Preheat Temp °C	120	145	120	145	130
Radiation (Heat Flux)	Similar				
Flame Length Visual m	1.8	1.5	2.25	2.00	1.25

However, in order to accommodate the larger coal particles in the coal mixture it was necessary to use a modified form of the internal mix burner atomizer tip. It is expected that similar modification would be required to burn additive stabilised mixtures. This is because additives do not alter coal particle sizes but only improve storage stability.

Although good combustion was achieved for both fuels, burner parameters were not so critical for satisfactory burning of the dispersion as they were for the coarser mixture. This would prove beneficial in installations where burner parameters are not so easily adjusted. In general the dispersion handled like a fuel oil whereas the mixture clearly required more careful handling.

Combustion Test in a Marine Boiler Aboard the MV British Hazel

Some 1200 gallons of the stable dispersion were burned in a BABCOCK WILCOX auxiliary boiler (a single pass water tube marine boiler) on the above tanker serving in the BP fleet. The boiler was fitted with two standard WALLSEND steam-assisted atomisers which were used without any modifications. A schematic diagram of the boiler is shown in Figure 10.

Two combustion tests were performed, one with a BP coal dispersion and the other with heavy fuel oil drawn from the ship's bunkers.

The overall impression gained during the firing of the COD in the auxiliary boiler was very encouraging. 4680 kg (1040 gallons) of dispersion were successfully burned without interruption at a rate of approximately 465 litres/h (105 gallons/h).

Ash components analyses on the combustion chamber deposit samples after the COD burn indicated a high degree of 'burn out'. The weight of ash collected from the combustion chamber was 8.0 kg which accounted for 9.6 per cent of the total ash input. No slagging occurred during the coal/fuel oil dispersion test; there was evidence of glazing around one burner quarl, but it is thought that this occurred whilst

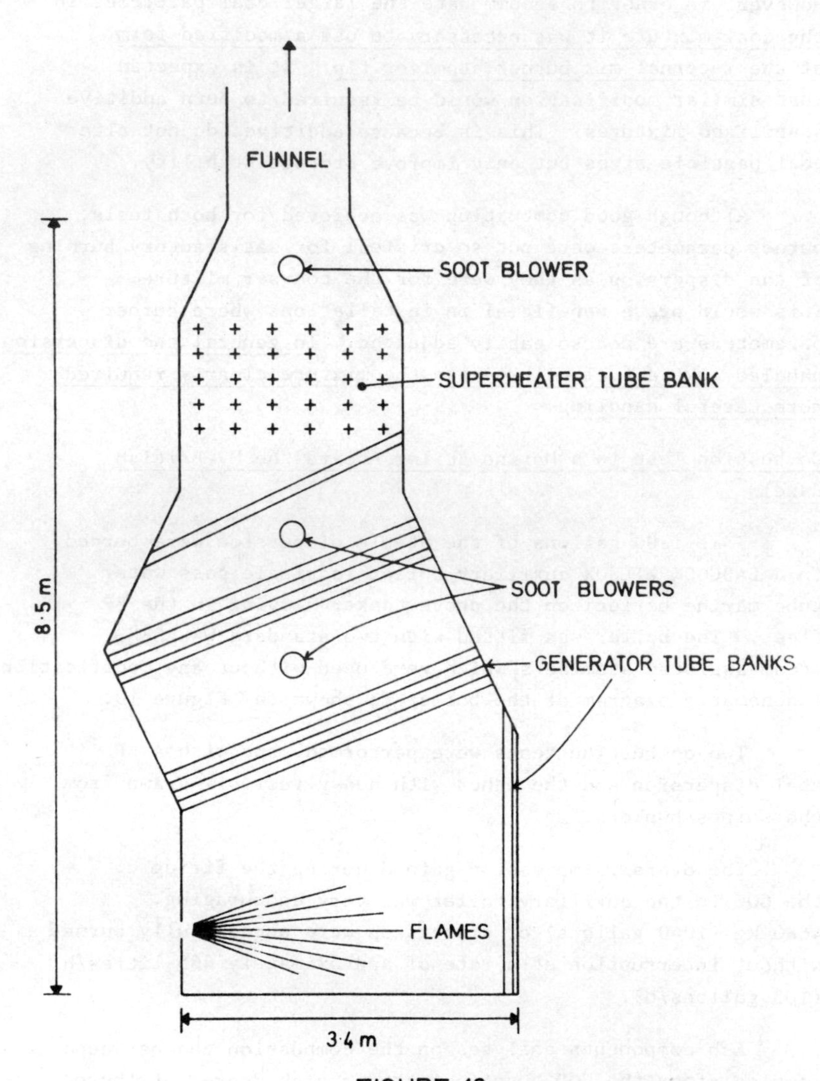

FIGURE 10
SCHEMATIC DIAGRAM OF BABCOCK AND WILCOX SINGLE PASS WATER TUBE MARINE BOILER

firing heavy fuel oil (ie before this trial), when flame inpingement was observed.

With this small amount of fall out, ash deposition was not seen as a serious problem in this type of boiler, where only the rear wall has heat exchange surfaces. However, only a longer term trial would determine any effect of tube deposition on heat transfer and how often the boiler would need cleaning.

It was estimated from an ash balance that 37.6 kg of ash were deposited on the 190 m^2 of generation and superheater tube surfaces, accounting for 44.6 per cent of the total ash input. There was no significant increase in uptake temperature or decrease in superheated steam temperature during the 10 hour burn, indicating that the <u>tube deposition was insignificant</u> and that heat transfer had not deteriorated.

The small tube deposits were easily removed by the sootblowers. This was not surprising since the superheat temperature was well below the fusion temperature of the ash contained in the dispersion (1230°C) and only a very light friable coating of deposit had formed.

The test served to establish the stability of our dispersion under conditions existing at sea, ie cyclic motion due to the waves as well as gravity forces. The dispersion was held in lagged tanks heated by steam coils nominally to 50°C, but in practice the temperature variation ranged from 80°C to 110°C. In spite of this, however, the dispersion was burned and the test completed satisfactorily.

Future Activities

BP is now in the process of constructing a plant suitable for the production of 100 000 tons per annum of stable coal/oil dispersions. It is expected that the plant will start to operate early in 1982 and produce sufficiently large quantities of the dispersion for long-term tests in major utility installations and major industrial users of fuel oil. Such long-term tests are essential to establish the effect of the dispersion on wear and erosion of equipment normally used for the combustion of fuel oil, as well as the general behaviour of the large quantities of the product in large-scale transportation and storage. All these tests are expected to be completed by the end of 1983, by which time we shall also have a definitive idea about the cost of production and distribution of the dispersion on a multi-million ton per annum scale to the major world markets. We expect that this new fuel will have an important role to play in the next 10 - 15 years and that it will ease the financial stresses that will be imposed onto the industry during the transition from oil to coal as a major source of energy.

Conclusions

BP has developed a novel method for producing very stable coal/fuel oil dispersions. Such coal-containing fuels have been handled in systems designed for heavy fuel oil. Quantities of coal in fuel oil dispersion have been prepared, handled, stored and fired; no coal deposition problems were encountered, even though occasionally large batches of the fuel were held in storage at temperatures exceeding 100°C for several days.

The production of a stable dispersion offers the possibility of using centralised production facilities, the potential for utilisation of current heavy fuel oil distribution infrastructure, and an economically attractive route to converting coal into liquid energy.

Large-scale, long-term trials are being organised by BP to demonstrate the feasibility of handling and burning such coal-containing fuel in boilers, blast furnaces, etc, designed to operate on heavy fuel oil and, in a few cases, coal.

Acknowledgement

Permission to publish this paper has been given by The British Petroleum Company Limited.

References

[1]G. Whittingham, J. Winsor, Development of a Fuel-Oil-Slurry Injection System for Blast Furnaces, Current Developments in Fuel Utilisation, Inst Fuel Brussels Conference, 7th - 11th September 1964.

[2]G.D. Botsaris, Y.M. Glazman, M. Adams-Viola, P.S. Goldsmith and R.J. Haber, Characterisation and Structural Studies of the Various Types of COM, Proc of 2nd Int Symp on Coal Oil Mixture Combustion, November 27th - 29th 1979 US Dept of Energy.

[3]A. Grint and D.L. Perry, to be published

[4]A. Naka, The Use of Additives to Stabilize Coal Oil Mixtures, Proc of 2nd Int Symp on Coal Oil Mixture Combustion, November 27th - 29th, US Dept of Energy.

[5]M. Hatano, Technical Results of EPDC's COM R and D. Step 1. Laboratory Tests, EPDC (Japan).

Technology and Chemistry in British Gas: Evolution and Revolution

By J. A. Gray, O.B.E.[1] and F. E. Shephard[2]
BRITISH GAS CORPORATION, RESEARCH AND DEVELOPMENT DIVISION
[1]NATIONAL WESTMINSTER HOUSE, 326 HIGH HOLBORN, LONDON WCIV 7PT, U.K.
[2]LONDON RESEARCH STATION, MICHAEL ROAD, FULHAM, LONDON SW6 2AD, U.K.

1. Introduction

The gas industry in this country was made possible through some early experiments by inquisitive scientists three centuries ago and, since that time, much of the industry's success has been based on a fruitful combination of ideas and expertise from a wide variety of scientific disciplines such as engineering, chemistry, physics and mathematics. This dependence on ideas and their implementation, together with the ability to communicate and co-operate towards the solution of problems, which was so important at each stage in the historical development of gas as a major form of energy, is still essential for further progress. It is a tribute to the quality and flexibility of those who have worked in the industry in the last twenty years that it has successfully come through two technological revolutions. Firstly from an industry which was based on the manufacture of gas from coal to one in which the bulk of gas was produced from liquid hydrocarbon feedstock, and secondly from that based on steam reforming and hydrogenation of petroleum fractions to one based entirely on natural gas. The present natural gas reserves will last well into the next century but already it is clear that methods for manufacturing gas will eventually return in some form although we cannot exclude unimagined advances in exploration techniques as well as in the methods of winning new sources. British Gas is already anticipating that demand in the long term and has been studying processes which will be able to provide substitute natural gas (SNG) when the need arises. Indeed some aspects of these developments have attracted substantial financial participation from abroad.

It is worth noting that the structure and organisation of the gas industry have also changed several times in the last few decades. The industry was nationalised in 1949 and in the place of the one thousand gas manufacturing works in operation at that time twelve autonomous area boards were established. The central Gas Council had three main responsibilities: for research, industrial relations, and capital expenditure. By the Gas Act 1972, the British Gas Corporation was formed on 1st January 1973 to replace the Gas Council and the twelve Area Boards. Thus although there was a continuous history, in recent times the name of the Industry and the organisation of its various component parts have changed.

This paper reviews the development of the gas industry from the days of coal carbonisation to the present day and looks to the needs beyond, and indicates briefly how problems or opportunites are met by expertise from within and without the Industry.

2. The Past - The Supply of Town Gas

2.1 Gas from Coal

Almost a century ago gas undertakings had an annual sale of 75 billion ft^3 of gas and the gas being sold was already being independently tested for purity and pressure. Lighting was the first main use for gas but, although its heating value was soon realised, it was not until 1920 that this use was actually reflected in the unit used to determine the quality. In that year there was a change from candle power to the therm. (1 therm = 10^5 Btu = 2.52×10^4 kcal = 105.5 MJ).

Town gas, as it was later called, was produced at first in horizontal retorts which were charged and discharged by hand as a batch operation but eventually those designs were replaced by vertical retorts which were usually continuous in operation. Coal was fed in at the top and coke discharged at the bottom. By that time the chemical processes which could be built onto the plant to clean up the gas had made substantial progress too.

In addition to coke, there was a wide range of useful materials which could be obtained directly or indirectly as by-products. They included benzole motor fuel, sulphuric acid, fertilizers, ammonia, refrigerants, explosives from the sulphur and ammoniacal liquor. From the tar, other industries could separate or synthesise a very diverse group of substances:- plastics, naphthalene, vitamins, insecticides, paints, dyes, nylon, medicines, perfume, wood preservative, disinfectants and road tar.

The production of coke had its own disadvantages, so some early attempts were made at the total gasification of coal. As suitable coking coals became more expensive, the Industry looked towards gasification methods such as the Lurgi process which offered gas production at pressure. The coal is reacted with oxygen and steam at high pressures and temperatures and the only solid residue from the reactor is ash. The gas produced has a calorific value (400 Btu/ft^3, 3572 kcal/m^3) lower than some distributed gases but the Lurgi system is superior to the previous continuous low pressure processes. After considerable development work had been carried out by the British gas industry to adapt the process, two plants were built, one at Westfield in Fifeshire and one at Coleshill in the West Midlands. In operation these plants produced large quantities of weak ammoniacal liquor because of the large steam requirement. The characterisation and treatment of that liquor was an additional aspect which required study for the successful operation of these plants. But already as the first of two major plants came on stream in 1961 the days of that coal gasification period were numbered. The change would result in a further large reduction in the number of manufacturing sites which had come into public ownership in 1949.

The need to dismantle old plant, and sell surplus sites, meant that attention was necessary to ensure that these operations were carried out in accordance with regulations in force at the time. That meant for instance devising procedures for sampling lagging, identifying any forms of asbestos present and gaining expertise in sampling and counting airborne fibres. It also entailed devising procedures for the acceptable disposal of any bulk residues from previous operations. The disposals included those of holder water, tar emulsion, spent oxide which had been used for gas purification, and surplus catalyst. Some of the sites involved had been producing gas for over a century and the presence of some residual contamination or unknown underground tankage and pipework could not be ruled out. However many of those sites have now been successfully redeveloped for other uses.

2.3 Gas from Oil

In the early sixties several factors led to an expansion of the gas industry at the same time as the Clean Air Act was being implemented. The Gas Council was responsible for the industry's research programme and the integration which followed nationalisation provided the impetus which was to yield the first major technical change by way of the evolution of new processes for gas manufacture. The processes flowed from the research on the total gasification of coal and of hydrocarbon feedstocks carried out initially at the University of Leeds, then by the pre-1949 Gas Research Board,and finally at the Gas Council's own research establishments.

Over the same period in which the Clean Air Act was being implemented, the price of oil products was falling steadily and provided a market opportunity to make gas in a more economical way. The first advances in steam reforming were by processes in which hydrocarbons were reacted with steam over catalysts, sometimes in cyclic plants similar to those used in water gas production. These processes produced gas at low pressure but they were soon superseded by the ICI, Topsøe and other continuous high pressure processes in reforming plants that used cheap light petroleum distillate feedstock and more advanced catalyst formulations.

The processes had all the advantages of pressure, large unit size high efficiency, the facility for carbon monoxide conversion and subsequent carbon dioxide removal. The pressure at which the gas was generated was compatible with distribution into Area Board high pressure grids. The processes were the first to achieve these advantages without using oxygen and the final gas produced was cheaper and far less toxic than the old town gas.

The simplest feedstock used was methane and in that case reforming involves two equilibria

$$CH_4 + H_2O \rightleftharpoons CO + 3H_2 \quad \text{endothermic}$$

$$CO + H_2O \rightleftharpoons CO_2 + H_2 \quad \text{exothermic}$$

In the first reaction an increase in temperature and a greater proportion of steam will push the reaction to the right whereas an increase in pressure will tend to push the reaction to the left. In the second reaction pressure has no effect but an increase in steam would push the reaction to the right and increasing the temperature would push it to the left. So the steam reforming of hydrocarbon was basically a mechanism of achieving the thermodynamic mixture of carbon monoxide, carbon dioxide, hydrogen, methane and steam at specific conditions. The actual amounts obviously depended on the type of feedstock as well as temperature, pressure and steam/feedstock rates. Figure 1 shows the variation in the equilibrium mixture with methane as feedstock and a weight ratio of steam to methane of 1.6 and a pressure of 25 bar.

Higher hydrocarbon fractions were the more usual feedstocks, and although the equilibria were essentially governed by the same equations as given above it was necessary to consider the breakdown of the higher hydrocarbon with steam. For example with hexane as feedstock the reaction could be considered as

$$C_6H_{14} + 2.5\ H_2O \longrightarrow 4.75\ CH_4 + 1.25\ CO_2$$

It was important to consider another reaction: the Boudouard equilibrium

$$2\ CO \rightleftharpoons CO_2 + C$$

Any build up of excess carbon monoxide could lead to the very rapid laydown of solid carbon. To prevent this undesirable side reaction the steam/hydrocarbon ratio had to be kept above a minimum value determined by the nature of the hydrocarbon and the operating conditions.

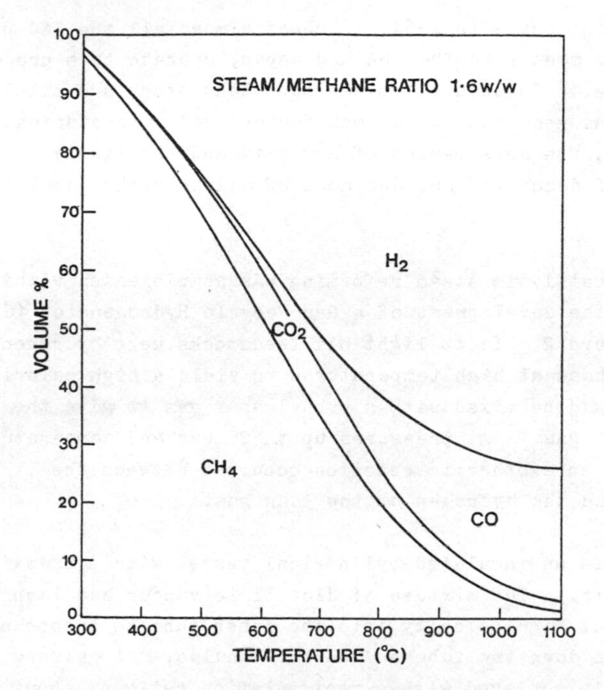

Figure 1 Variation in equilibrium gas mixture with reaction temperaturo

In order to supply town gas with a calorific value of about 500 Btu/ft^3 a variety of process routes became available and, although the gas industry developed some robust catalysts for "lean" gas production, the main effort involving catalysis was concentrated on the Gas Council Catalytic Rich Gas (CRG) process. Compared with the ICI process it operated at a low temperature to give an equilibrium product high in methane and thus required a very active catalyst which would operate at a low temperature and also resist sintering and carbon formation and produce a gas that could be distributed without enrichment. An effective desulphurisation unit was also required to pretreat the reactants and thus to protect the reforming catalyst from sulphur poisoning.

A CRG nickel alumina catalyst was developed and the process has achieved success both at home and abroad since its development to a commercial scale in 1964. Indeed almost all the SNG plants in the world, mostly in the USA and Japan, operate this process under licence. To ensure success much laboratory and pilot plant work was required on the manufacture and characterisation of catalysts, the measurement of activity and the likely mechanisms of deactivation, and most of all on strict quality control.

The work on catalytic steam reforming was complemented within the Industry by the development of a Gas Recycle Hydrogenator (GRH) shown in Figure 2. In it light oil feedstocks were hydrogenated in the gas phase at high temperatures to yield a high calorific gas which could be mixed with a much leaner gas to give the required town gas. At pressures up to 25 bar and temperatures of 700-750°C an exothermic reaction occurred between the distillate and the hydrogen of the lean gas.

The reactor is an insulated cylindrical vessel with a co-axial tube inside it. The mixture of distillate vapour and lean gas is injected at high velocity into the inner tube to induce a recirculation down the tube and up the annulus. A uniform temperature is achieved with a recirculation ratio of about 10 to 1. The methane and ethane formed came essentially from the aliphatic content of the feedstocks and the aromatic nuclei which survived the reaction could be used as a source of high-purity benzene. Research studies of the homogeneous rections of hydrocarbons yielded greater understanding of the reaction sequences involved in the GRH reactor, and how to optimise the reaction conditions.

This whole period of expansion, involving the use of hydrocarbon feedstocks, was based on a mixture of sound understanding of thermodynamics, kinetics, process engineering and catalyst characterisation and the art of enthusiastic scientists and chemical engineers. The gas industry had to develop special expertise in handling these plants so that they could respond to variable demand conditions. The cycling of high and low output demanded fresh developments in design and operating conditions of plant materials. It was a short lived period but one, as we shall see, which laid a good foundation for future processes.

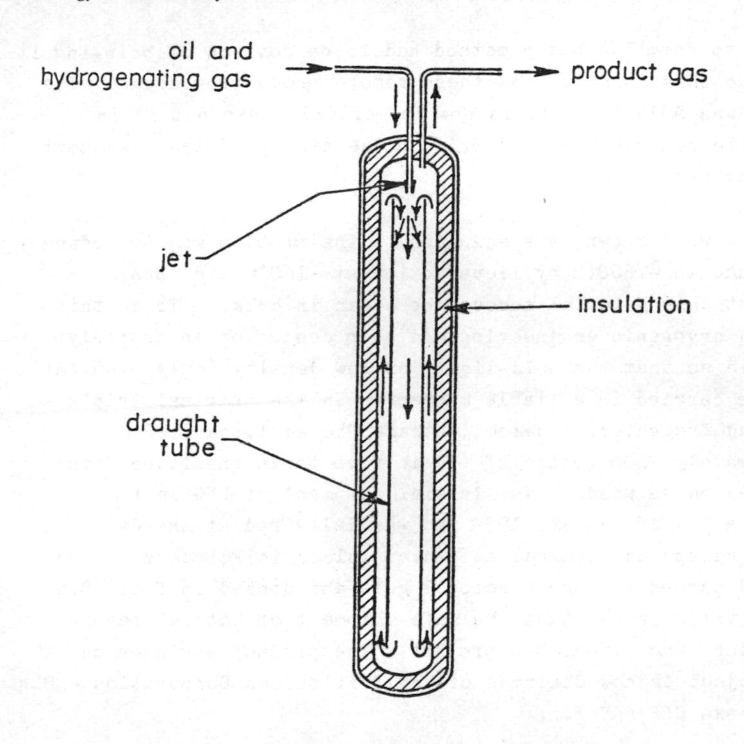

Figure 2 The Gas Recycle Hydrogenator

3. The Present - the Supply of Natural Gas

3.1 The Introduction of Liquified Natural Gas

The first significant development towards the supply of natural gas on a national scale was in 1961 when the Government gave approval to the Gas Council's request to import liquid natural gas (LNG) from Algeria over an initial period of fifteen years. It was the first project of its kind in the world and arose from Gas Council interest in the large quantities of natural gas available overseas but which, at that time, were being flared in areas remote from markets and pipelines. The gas could be chilled to form LNG but a method had to be devised of bringing it by sea to this country from these remote production areas. The problem was solved and it is now in worldwide use but it is worthwhile remembering that British Gas pioneered the transport of LNG by sea.

As is now well known, the solution to the problem was to reduce the volume to 1/600th by liquefaction at -160°C and then transport this material across the ocean in bulk. To do this required cryogenic engineering and ship design on an entirely new scale so that the cold liquid of low density (only 0.45g/ml) could be carried in a stable manner. In the original trials a converted freighter, renamed 'Methane Pioneer', carried approximately 2200 tonnes of LNG in five tanks insulated with laminated balsa wood. The initial shipment of LNG left Louisiana USA in January 1959 and was delivered at the Gas Council reception terminal at Canvey Island in February. The ship had passed through a Force 9 gale and docked in fog! Six further trips proved that the safe shipment of natural gas in liquid form was a feasible project. The project engineer on that project is now Chairman of the British Gas Corporation - Sir Denis Rooke CBE FRS FEng.

For the main operation, importation from Algeria, two ocean-going LNG carriers, the 'Methane Princess' and the 'Methane Progress', were built in the UK and launched in 1963. The ships both had a capacity of about 12,000 tonnes (approx 27,400m^3) of LNG and in size were equivalent to a conventional oil tanker of about 28,000 tonnes. The first commercial cargo was delivered by the Methane Princess in October 1964. This operation of importing LNG necessitated the construction of the first large high pressure transmission line, that between Canvey Island and Leeds, with supply spurs to eight Area Boards. For some time methane was used to enrich reformer gas to a calorific value of 500 Btu. This first backbone main provided the experience on which the present national high pressure transmission system was built. The discovery of natural gas at Groningen in Holland, and at Whitby in North Yorkshire, was already giving hope that natural gas might be discovered in the North Sea.

British Gas has participated in the offshore activity from the earliest days in one of the consortia granted licences, in 1964, for exploration in the UK sector. Without this foresight British Gas would not be as successful as it is today. Its Exploration Companies have been responsible for the discovery and production of oil from the largest on-land development in the UK - Wytch Farm in Dorset - and for the gas field in Morecambe Bay in which British Gas will is the sole Operator and about which more is said later.

3.2 Conversion from Town Gas to Natural Gas

The sales of gas in the 1960s increased rapidly even before natural gas finds in the North Sea. In the short term it was possible to reform natural gas and distribute the resulting town gas but, with the prospect of larger markets two factors about future policy became important. If the increased demand was to be met through the installation of further reforming plant both a large capital investment and a commitment to considerable running costs were inevitable. At the same time storage facilities would have had to increase enormously. The prospect of changing to an industry based on natural gas had the attraction that it would effectively double the storage,transmission and distribution capacity purely by virtue of its higher calorific value (approximately 1000 Btu/ft^3). The main difficulty with such a changeover was that it would require the conversion of tens of millions of appliances over a period of years in order that natural gas could be used efficiently.

Because natural gas requires about double the air per unit volume that town gas required, burning of natural gas in an unaltered town gas burner would have resulted in a floppy, sooty flame. Even if the air added was increased for the same volume throughput of gas the flame would have been too large for the appliance as would the heat output. Thus conversion required the supply of the same heat in a given time and a flame of the same size and shape as before. These problems were overcome through the use of small injectors for natural gas and by increasing the pressure of supply. Flame speed is also an important factor; it is the rate at which a flame is propagated through a mixture of gas and air. Natural gas has a slower flame speed than town gas, and burner design had to be altered to prevent the flame 'lifting off' the burner.

This whole area of burner design, its dependence on the gaseous mixture being supplied to it, and the performance under varying conditions, had been studied in depth for many years. The experience gained was critical in providing the technical advice and testing in the time leading up to and during conversion.

In June 1966 the then Chairman of the Gas Council, Sir Henry Jones, announced the decision that was to revolutionise the industry: this was to convert the whole country to natural gas in the ten years 1967 to 1977.

A pilot scheme was carried out at Canvey Island, supplied from LNG storage, in 1966 and it demonstrated the technical and administrative feasibility of large scale conversion in a way that would be acceptable to the customers. In all, the conversion took about ten years to complete but within six years 73% of customers had been converted. When it finished in 1977, 13.5 million premises had been visited at least twice, 35 million appliances were converted, including 200 million burners and 8000 different domestic appliances. In real terms, the cost was less than that estimated at £400 million in 1966. This was an operation on a scale and at a speed that has not been matched anywhere else in the world.

3.3 The Build-up of Natural Gas Supply

In the Southern Basin the discovery of commercial quantities of natural gas in the British Sector of the North Sea in 1965 gave fresh impetus to an industry that was already changing radically through the introduction of processes based on petroleum feedstocks. One year after the decision was taken in 1966 to convert from town gas, natural gas was already coming ashore from the British Petroleum field in 1967 at the first terminal at Easington. Further large scale discoveries in the southern North Sea were made by consortia led by Shell/Esso; by Phillips; and by the Amoco group in which British Gas participated, and led to the construction of a large gas reception terminal at Bacton; it was completed in just over a year and was commissioned in August 1968. The third shore terminal at Theddlethorpe became operational in 1972. In the Northern Basin the Frigg field was discovered by the Petroward Group in 1971 and the first gas from this field was received in 1977 at the St. Fergus terminal which was officially opened by Her Majesty the Queen in 1978 (Plate 1).* As natural gas gradually replaced town gas the average daily gas available increased as shown in Figure 3 from just less than 10

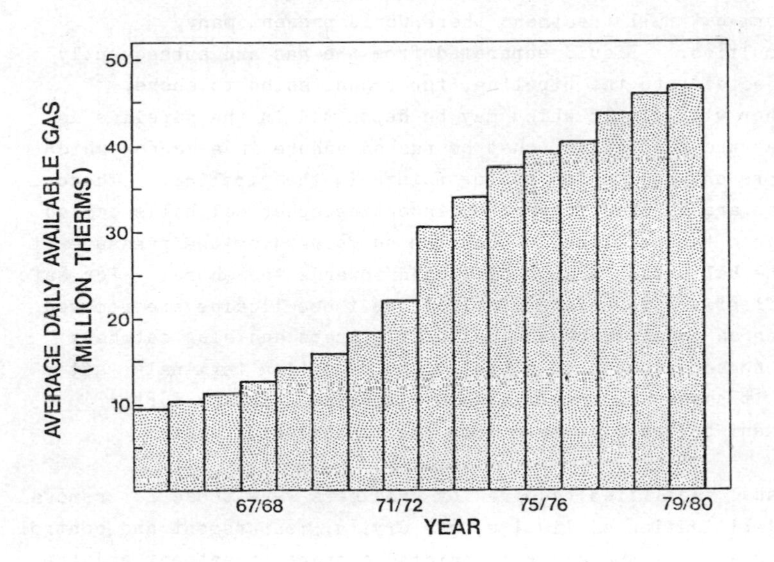

Figure 3 Diagram showing the average daily gas available

to more than 45 million therms (ie about 4500 million ft^3 a day or 1230 million m^3 a day) over a 15 year period.

From the beginning of the natural gas era, the industry was well aware of the environmental dangers and objections which would be raised with respect to the siting in rural areas of onshore terminals and all the other projects necessary to develop the full potential of North Sea gas. In order to minimise costs it has been the policy of the Industry to establish a limited number of coastal reception terminals. This approach also minimises

*Plates are on pp. 118 and 119.

any effects on amenity. At present four terminals receive gas from the fields referred to above i.e. West Sole, Rough, Viking, Hewett, Indefatigable, Leman and Frigg. A fifth terminal is under construction at Barrow in Furness to receive gas from the Morecambe field (referred to later).

Natural gas cannot be transferred straight from the well to the transmission network because it contains solids, liquid hydrocarbons and water which must be removed. Some treatment to remove easily separated impurities is possible on the offshore platform but full treatment there would present many difficulties. Liquid separated from the gas and subsequently re-injected into the pipeline, for transmission to shore, together with liquid which may be deposited in the pipeline as the gas pressure falls, must be pushed ashore in a manner which will prevent collection at low points in the pipeline. These liquids are carried forward by inserting spherical balls called pigs into the pipeline. When pushed forward by the pressure of gas the balls push a slug of liquid towards the shore. For safe and effective operations of pipelines these liquids are removed as soon as possible by the use of pig traps and slug catchers (nice nomenclature!) on-shore at the reception terminals, after which it passes to the British Gas terminal for any further treatment before it passes into the transmission system.

The usual facilities required by producers were those for removal and stabilisation of liquids, gas drying, measurement and control of flow and all the other associated storage, treatment and the necessary offices and workshops. Indeed there were often a number of producers bringing pipelines ashore from different fields to the same area of coastline.

The transmission system to receive and transport natural gas from the terminals grew rapidly. It was almost 700 miles long in 1969, over 2200 miles in 1973 with four compressor stations, over 3000 miles and eleven compressor stations in 1979 and in the 1980's it is planned to expand to 3700 miles with twenty-two compressor stations sites.

The design of this high-pressure system ensures that the maximum throughput can be achieved from sources of supply to areas of demand. Some of the gas in the system is available to meet short term demands. There are other advantages. There is minimal disturbance to the environment: land through which the pipelines are laid can be completely restored and the supply is not vulnerable to the weather. Great care was taken at the design stage to ensure that shore terminals and compressor stations have the best visual impact and that all necessary noise control measures for the individual locations are taken.

3.4 The Challenges of Natural Gas - Some Solutions

3.4.1 Composition and Properties

Before the first gas was brought ashore much work had been done to ensure that the quality of gas received after treatment would be satisfactory for transmission and use. It was important to build up the analytical expertise and accumulate data on the various natural gases likely to become available, not only because the calorific value and so purchase price depended on composition, but because the combustion characteristics were also important to those involved in designing and approving appliances. The full analysis of natural gas was obviously already possible by gas chromatographic methods but further methods were developed which would give rapid quantitative analysis of complex gas mixtures of compositions similar to those shown in Table 1. The composition of a re-evaporated LNG is also included for comparison purposes.

One aspect of the gas composition which was of particular interest was the class of compounds which would give an odour. Distributed town gas had an odour which most people recognised and it served as an indicator of unlit or leaking gas. Some of the natural gases had traces of organic sulphides and mercaptans which have very strong odours even in very small quantities. Interest therefore focused on developing a method for quantitative determination of those compounds, at parts per million levels and lower, in the presence of hydrocarbons. No commercial instrument was available so a system was developed using flame photometric detection of excited S_2 species. The

Table 1 Composition of three natural gases and an evaporated LNG (in mole %)

Constituent	Gas A	Gas B	Gas C	Evap.LNG
Nitrogen	0.52	1.16	1.22	0.36
Helium	0.004	0.03	0.03	nil
Carbon dioxide	0.32	0.48	0.04	nil
Methane	95.69	94.02	94.81	87.20
Ethane	3.40	3.25	3.00	8.61
Propane	0.03	0.61	0.55	2.74
iso-Butane	0.01	0.11	0.09	0.42
n-Butane	0.01	0.13	0.10	0.65
Pentanes	0.01	0.18	0.06	0.02
Hexanes	0.01	0.04	0.03	nil
Heptanes	< 0.005	0.02	0.02	nil
Octanes	< 0.005	0.02	0.01	nil
Nonanes	< 0.005	< 0.005	< 0.005	nil
Benzene	0.005	0.04	0.03	nil
Toluene	0.005	0.01	0.01	nil

final instrument analysed quantitatively for methyl mercaptan, t-butyl mercaptan, ethyl mercaptan and disulphides in less than twenty minutes. It was then possible to develop mixtures which, when added to natural gases in small quantities, gave suitable odour intensity and impact so that the odour experienced by the customer would be recognisable and constant. A typical analysis of the sulphur compounds in a distributed natural gas is shown in Figure 4.

The next stage was to construct the instrument in an automated form to provide on-line data on the odorant composition of the gas leaving the terminal. Such an instrument has been operating virtually non-stop for several years at the Bacton terminal and it is now in commercial production.

In considering the processes necessary for producing, treating, transmitting and using natural gas there are problems involving

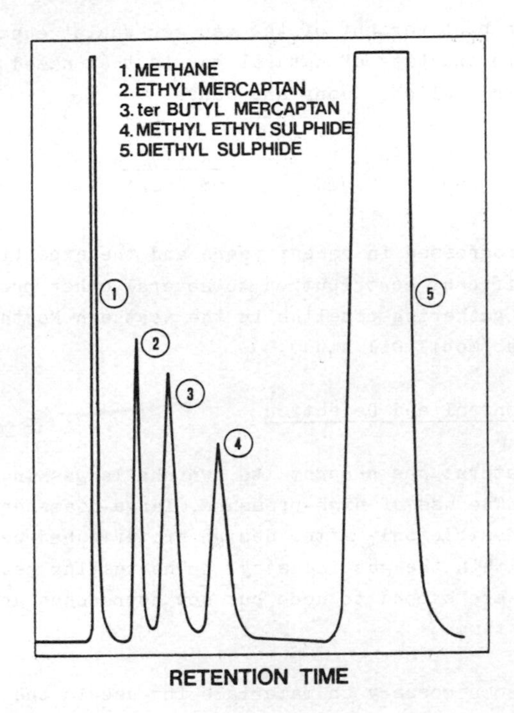

Figure 4 A typical chromatogram of organic sulphur-containing compounds in natural gas

thermodynamics: data were already available for pure substances but not for natural gas mixtures which contain quantities of several other gases as we have seen in Table 1. Theoretical models were available which might be used to derive the behaviour of natural gas and an assessment of them showed, perhaps surprisingly, that the simple van der Waals' equation

$$\frac{PV}{RT} = \frac{V}{V-b} - \frac{a}{RTV}$$

was more appropriate than many modern derivations with many

constants. The best variant of the van der Waals' equation for calculating the properties of natural gas is that named after its modern developers Redlich, Kwong and Soave.

$$\frac{PV}{RT} = \frac{V}{V-b} - \frac{a\,(T)}{RT\,(V+b)}$$

This work has progressed in recent years and the expertise gained has made a significant contribution to several other projects e.g. to the gas gathering pipeline in the northern North Sea - a joint British Gas/Mobil Oil study.

3.4.2 Leakage Control and Detection

The advent of natural gas brought its own challenges in the field of materials. The use of high-pressure, large-diameter pipelines was possible only after design criteria had been drawn up by experts within the gas industry. The testing methods and facilities used are second to none but new approaches are being studied all the time.

Research was also necessary on materials for use in the lower-pressure distribution systems. Many of the joints in the traditional distribution system depended on the swelling properties of liquids present in coal gas on joint packing to ensure an effective seal. The replacement of coal gas by town gas made from gas or oil feedstock and subsequently by natural gas made the 'drying out' of these joints more significant.

The solution for the joints with jute as the sealing medium depended on the swelling properties of jute yarns and was a treatment of mains with a liquid such as monoethylene glycol which entered the jute structure, causing it to swell and thus keep the joints gas-tight.

An important influence in leakage control is the steady replacement of older materials by plastics. Where joints between and in metal systems required sealing both external (encapsulation) and internal systems were investigated.

Medium density polyethylene (MDPE) has proved to be the most satisfactory material for low-pressure distribution systems below 4 bar. Its acceptance was dependent not only on its meeting technical criteria, but it also had to be competitive with metal systems in economic and commercial terms.

A modern distribution system must withstand internal gas pressure and external stresses as a result of exposure to the surrounding soil; also the effects of interference from third parties must be minimised. Although MDPE does not corrode, it is susceptible to creep at ambient temperatures so any design must give an acceptable system life before any deformation produces a risk of failure.

Studies into the fundamental behaviour of MDPE have been wide ranging. Particular attention has been directed to determining those parameters which govern the slow growth of stress cracks, the most likely long term failure mechanism, and a study of the factors involved in high speed fractures. The ductility, flexibility and fracture toughness properties of MDPE have been shown to offer a system capable of safe long-term operation. A particular advantage of its flexibility is that it can be bent, within specified limits, in continuous lengths and this avoids the introduction of large numbers of joints.

A polymeric material such as MDPE may be influenced by organic materials and other chemical substances sometimes in the gas, and the soil environment in which it has been laid. Therefore work was undertaken to ensure that none of these situations could jeopardise long and safe life.

The growth in the use of this polymer material has been spectacular and involves a consumption of some 15,000 tonnes each year. By April 1979 nearly 14,000 km of mains and 2.75 million service pipes from the street to the dwelling had been laid in MDPE. It now represents more than 70% of all new and replacement mains and services. Work is still in progress to extend operating conditions and optimise further use.

Preventive maintenance of the gas distribution system is a

continuous operation but in a system with 60 million joints it is not surprising that some leakage occurs and it is essential to detect such leakages quickly and reliably. For this the Industry has designed special road vehicles with equipment which can detect extremely low concentrations of gas. These patrol distribution mains at regular intervals and, when a leak is indicated, the next step is the use of a portable detector to pinpoint it. In recent years a special instrument, the 'Gascoseeker', has been developed to meet the particular needs of the gas industry. It had to be robust, reliable, sensitive and operate over a wide range of gas concentration. In the final design chosen there is a filtering system to remove dust and any constituent likely to poison the detector element. Two sensors are used to give readings as a percentage of the lower explosive limit (LEL). The design has resulted in an intrinsically safe instrument suitable for use in hazardous atmospheres and in most field operations. Thousands of the instruments are now in use.

Higher pressure distribution pipelines require further treatment to prevent leakage. This can be done either internally or externally. In order to assess the methods proposed it was necessary to build up experience in the formation, properties and analysis of polymers such as polyurethanes. The properties of fluid sealants were also studied. This background information led to a successful solution to the problem of sealing leaks in the service risers which carry gas from street level to the individual floors in multi-storey buildings. Often the risers are located in ducts which make access difficult.

The method chosen was a fill and drain operation in which the chemical used was developed from several years of fundamental work by British Gas on the formulation and application of thick coating sealants. The sealant had to meet exacting requirements. It had to be completely safe, cheap and convenient to use. The final solution was a water based co-emulsion of bitumen and rubber with the addition of other specific components to give suitable properties.

3.5 Conservation in Utilisation

Despite the riches of energy available in the North Sea an energy conservation policy is essential. The Industry has a three pronged approached:- 1) "Don't Waste It"; 2) more effectient methods of utilisation; and 3) a sensible depletion policy. An example of the first has been the policy of replacement of pilot lights on cookers by spark ignition.

An important aspect of ensuring efficient use of the gas is to identify those potential applications where it has advantages over other fuels. This approach has led to significant design improvements which have in themselves resulted in energy savings in the industrial market. Three such developments were in the design of recuperative burners; in vat and tank heating; and in the rapid heating of metals.

The recuperative burners are designed to recover waste heat by preheating the combustion air. In the past their use has been confined mainly to continuous processes. British Gas has succeeded in designing a compact unit which can be mounted in the furnace wall with no more effort than a conventional burner as a burner itself. The waste gases are collected through an annular slot around the burner and preheat the combustion air in a concentric tube recuperator around the burner. Control systems can be built into the air flow for changes in temperature. The gas recuperative burners are economically attractive for batch operation. They are used for applications such as pottery firing, forging and metal melting and fuel savings of up to 50% have been achieved.

Vat and tank heating has been successfully carried out by passing steam through a simple U-tube immersed in the liquid, or by using direct gas fired natural draught immersion tubes. The former, although simple to install and operate, is usually less than 40% efficient because there are losses in generating and circulating the steam: the latter is efficient, usually about 70%, but occupies a large volume of useful tank space. The development by British Gas of a compact forced draught system, operated by direct firing of hot combustion products at high intensity down small bore immersion tubes, has offered the prospect of even higher efficiencies - say 80%. The heat input of burners has varied from 20 to 320 kW (0.07 to 1.1 million Btu/h). Many of them have been operating for several years.

Any technique that improves a metal reheating process by reducing the overall heating time may be labelled "rapid heating". With a clean high grade fuel such as natural gas, hot combustion products can be allowed to impinge directly on to the metal. Through the use of jets of hot gases from high velocity burners, and by tailoring the external shape of reheating furnaces to that of the stock being reheated, the convective heat transfer is increased. As a result the metal is rapidly heated and so energy consumption is lower than it might otherwise be. The application of this technology, some of it the subject of British Gas licences, has led to many sales, and a fair number of them have been overseas.

All the technical advances made in the design and operation of appliances obviously need to be implemented if the full benefit in energy saving is to be achieved. It was an appreciation of this need for education which led to the establishment of the British Gas School of Fuel Management at Solihull. There, alongside one of the Research Stations, British Gas staff and fuel managers from industry can be trained and updated in the efficient use of energy.

Much of the gas used in Britain, in fact more than 40%, serves to maintain a comfortable indoor environment. Conservation in this application requires not only information about the factors that affect efficiency and comfort such as ventilation, draughts, heat sources, air temperature, humidity and clothing worn within the building, but also on how the building itself interacts with outside weather conditions. A specially developed computer program, THERM, can simulate this behaviour by building in features such as air temperature, wind speed, solar radiation, shade from other buildings, the heating/cooling system, and the level of lighting. By altering the amount of insulation or double glazing, the thermostat settings and other such parameters on the final model, it is possible to see which conservation measures will give the best overall saving in energy.

In recent years there have been continued national attempts to persuade the general public to save energy. British Gas has given full support to these campaigns and a very comprehensive range of activities is undertaken to ensure that gas customers are not energy wasters. A unique aspect of these efforts was the establishment in 1979 of the West Midlands Gas Energy Advice Centre in Birmingham. It has extensive library, reference and study facilites and a computer is available for architects, students and planners to assess the thermal behaviour of structures.

One of the other ways in which British Gas seeks to highlight the vital need for the efficient use of energy is through sponsorship of two awards. The first are the Gas Energy Management (GEM) Awards. These are presented annually to partnerships between a gas user and a Regional Technical Consultancy Service which are judged to have made the most outstanding contribution to the efficient use of fuel during the year. In 1980, after five years of the Awards, it was estimated that the 258 entries to date had achieved savings of over 80 million therms a year - enough to serve a city the size of Bradford. The second Awards, inaugurated in 1979, are called the Design for Energy Management (DEM) Awards and are sponsored by British Gas in association with the Royal Institute of British Architects. These are promoted to encourage students of architecture to develop an understanding of the principles of energy efficiency in buildings.

3.6 Exploration Activities

Since the late 1930s there had been efforts to discover hydrocarbon deposits in Great Britain. Some finds were made: one of them at Eskdale in Yorkshire was used by North Eastern Gas Board from 1959. Even as far back as 1952 the Gas Council had invested in exploration activities by joining forces with British Petroleum Ltd to search for natural gas and also for structures underground suitable for the storage of peak load gas. The Gas Council entered into other off-shore exploration activities in partnership with major oil companies and on its own account.

The experience gained helped the industry to remain well informed and technically capable to anticipate developments. It has also resulted in practical successes; the expertise within the industry resulted in both major oil and gas finds - several new oilfields in Dorset (Plate 2) and the Morecambe Field in Morecambe Bay.

The Dorset oilfields are the largest on-shore oilfields in Great Britain and, although jointly owned with British Petroleum, British Gas are the operators. The Wytch Farm oilfield was discovered when the first exploration well struck oil at a depth of around 3000 ft. Further step-out wells proved the field was commercial. In 1977, our geologists developed a theory that a further oil bearing formation could lie beneath that first discovery. A deeper well was drilled from one of the existing Wytch Farm sites and confirmed the existence of what appears to be a larger oil discovery at about 5000 ft. In addition to the Wytch Farm oilfield several others have been discovered and a number of oil companies have submitted plans for exploratory drillings in adjacent areas including British Gas itself as sole licensee.

Reservoir assessment requires detailed information about the nature of the rock formations and their ability to store gas and flow gas and oil so that the reservoir can be modelled mathematically. Drilling cores are assessed for porosity and permeability which give basic information about the interconnected pore space in the reservoir rocks. By measuring the volume of mercury pushed into the pore system by increasing the applied pressure it is possible to estimate the distribution of pore size and the fluid flow behaviour. Further information can be obtained through optical and electron microscopy which supply information on the particle size distribution and morphology and phases present can be determined by X-ray diffraction.

It is equally important to collect information on the various fluids, both aqueous and organic, found in the structures. A typical reservoir at 2500m depth may have ambient conditions of 300 bar and 90°C with the result that fluids undergo significant changes during production including separation into more than one phase. One of the important determinations in studies on pressure-volume relationships is the 'bubble point' of the system that is the pressure at which a single phase system becomes two phase at reservoir temperature. Morever a full range of analytical techniques is necessary to describe adequately the samples taken during the drilling and assessment periods.

The fact that British Gas is the Operator means that it is responsible for the whole development including environmental protection. The area in which the oil has been found, Dorset, is considered by many as one of the most important ecological areas in Britain; it is a county with diverse geological strata and soils and therefore supports a diverse range of habitats. Of particular interest is the existence of acid heathland in very close proximity to chalk downland, together with the maritime influence afforded by the English Channel and Poole Harbour. This is the largest natural harbour in Europe and at high tide the water covers more than 4000 hectares. However, it is very shallow and at low tides extensive mudflats are exposed. The area is a favoured feeding and wintering ground for many wildfowl and waders so has great importance from an ecological point of view.

It is important that developments for gaining gas or oil should be carried out with the minimum disturbance to the environment. For the Dorset oilfields a monitoring programme was undertaken to ensure that British Gas activities did not adversely affect the environment. Baseline values were determined for selected parameters and regular assessment continues throughout the construction, drilling and production phases. In this way, information is built up about each site and any variations from the norm can be detected. The monitoring involves visual examination of each site, the collection of suitable samples and their preliminary examination on site. Samples are examined by several tests, some of which are selected because they are related directly to British Gas activities, eg. hydrocarbons in water, while others are more general measures of water quality or the presence of toxic substances. The magnitude of any deviation is assessed and corrective action recommended before

PLATE 1

PLATE 2

PLATE 3

PLATE 4

damage occurs. An example of some deviations in water quality at one site is shown in Figure 5. There are two large deviations in total solids concentration - in October 1979 and November 1980. During October it coincided with a high level of filterable solids while in the latter it coincided with a high value of sodium and chloride but the filterable solids remained low. It was easy with other information to confirm the interpretation that the

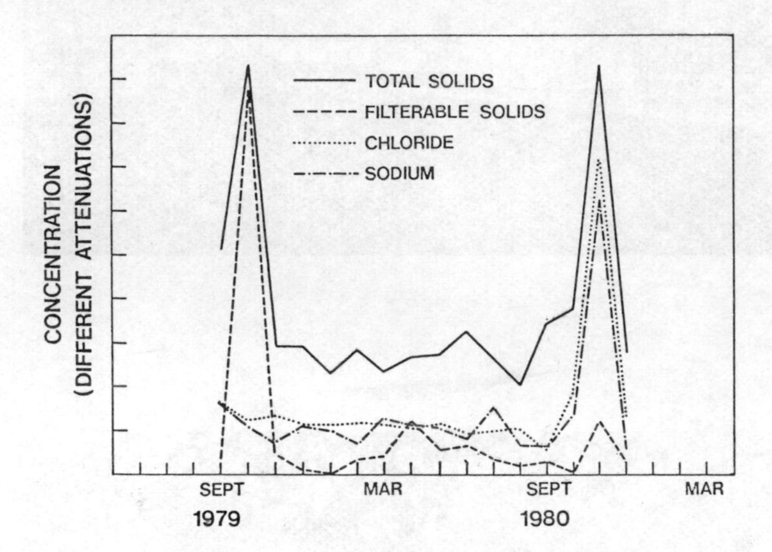

Figure 5 Changes in the concentration of total solids, filterable solids, chloride and sodium observed at a sampling point over a period of months

first peak was due to a construction period and the second to the presence of water from the lower formation levels which were being drilled at that time. Such assessments require not only a high standard of technical expertise, but also an accumulated bank of background information on environmental protection.

4. The Future - Ensuring Gas Supply

4.1 Supplies of Natural Gas

In order to ensure that our gas resources are managed to the best advantage it is necessary to project well into the future - to the end of the century and beyond. Many of the assumptions and forecasts made about the future may be incorrect but the industry has developed plans and policies which it hopes will be capable of meeting most eventualities.

The marketing and depletion policies determine where the gas should go and how fast reserves should be depleted. Marketing policy ensures that it provides a level of sales sufficient to finance the business and its continuity while at the same time ensuring that gas is not "oversold".

A considerable amount of gas sold for premium use supplies a temperature dependent market which results in periods of high and low demand and a marked "peak-load" problem; while in theory it might be possible to increase or decrease the flow from the wells supplying the gas according to demand, in fact other measures have to be taken. This is because contracts with suppliers for gas with a low load factor would be very expensive.

Traditionally the gasholder has been a buffer against diurnal demand, but growth in the amount of energy supplied as gas could not be met practically by the construction of large numbers of gasholders. Some variation in supply is met by changing the flow of natural gas into the system and through the use of unobtrusive high pressure storage vessels and nests of high pressure pipes which can be sited away from residential areas.

One element of the required storage is provided by the siting of liquefaction plants and large LNG tanks (about 20,000 tonnes capacity) at strategic sites in the transmission system (Plate 3). Thus gas is liquefied and stored for use in periods of prolonged very cold weather. Some of these tanks are already in existence and others are in construction to meet future needs. The experience gained through the importation of LNG in the sixties was obviously useful when planning this increased storage capacity. Like terminals and compressor stations however, the tanks often need to be constructed on green field sites and great care is necessary to ensure that the visual impact of a storage site is appropriate to its surroundings. At present there is a total storage capacity of about 170,000 tonnes and it is planned to expand this to 270,000 tonnes. (This represents 14 billion ft^3 of natural gas; in a severe winter this could be supplied over a period of twenty days at a maximum peak of 2100 million ft^3 a day.

One solution is to store gas underground. Indeed such a newly developed storage facility which does exactly that is sited at Hornsea in Yorkshire. The particular advantage at that location is that a thick layer of salt exists deep below the surface and it provides suitable conditions for making cavities which can then be filled with gas at high pressure, so that it can be transferred directly into the transmission system. Cavities have been leached out under carefully controlled conditions determined by prior studies and detailed examination of drilling cores taken through the structure. Optical, X-ray and chemical methods yielded information on the relative quantities of various salts present at different depths and their solubilities. That type of information together with procedures to control the leaching, and to measure progress, ensure that the correct final shape of cavity is achieved. In the summer, when demand is low, these cavities can be filled with gas. When high demand periods come along the gas can be taken out, given any necessary treatment, and transmitted to where it is required.

Even the extra storage as LNG and in salt cavities is not the complete solution for meeting the peak demand expected by the end of the 80's. However there are several other options available; two of them involve offshore gas fields.

Last year British Gas successfully negotiated with its partners in the Gas Council/Amoco Group to take over control of the partially depleted Rough gas field just east of Easington as a naturally occurring storage facility. In January of this year it was announced that, as part of a £340m programme, three new platforms will be built for the Rough Field, two for drilling and one for process equipment. The figures quoted at that time were that by 1984 gas injection would be started and by winter 1984/85 it could be yielding 400 million ft^3 a day and nearer 1 billion ft^3 a day by the next winter. That compares with 170 to 180 million ft^3 a day when the Rough Field was producing at full capacity in the late 1970s.

The second opportunity arises from the discovery of the Morecambe Field in 1973. It is a medium size field estimated at 5 to 6 trillion (million million) ft^3 of recoverable reserves. The total cost of developing the field, estimated earlier this year as £1 billion, might involve the construction of up to twelve platforms. It is expected to come on line by late 1983 or early 1984.

More recently there has been considerable effort towards ensuring that gas associated with oil in the northern North Sea is not wasted. A joint study, initiated by the Government and undertaken by British Gas and Mobil North Sea Limited, looked at the feasibility of schemes for collecting and bringing to shore gas from condensate fields, small accumulations together with gas that would otherwise be flared. Onshore and offshore work on such a gas gathering system is now proceeding. The Organising Group consisting of British Gas, British Petroleum and Mobil have formed a joint venture and already more than £8 million has been committed. Full financial arrangements have yet to be made but it is hoped that an interim company will soon replace the joint venture.

These are the main options available in balancing supply and demand for natural gas. It is expected that further significant finds of natural gas and oil will be made in the relatively large areas around our coasts not yet fully explored but ultimately natural gas reserves must decline. They will not decline suddenly but rather will decline over a considerable period and can be replaced systematically, when required, using new purpose-built SNG plants. Thus the long term strategy depends on the three pronged conservation programme described earlier and further exploration. Renewable sources are unlikely to supply more than 10% of the total energy needs by the year 2000 so when oil and gas supplies need to be augmented the balance will be between coal and nuclear power.

As far as the gas industry is concerned the supply of natural gas will eventually be augmented by SNG perhaps initially manufactured from oil and later from coal and in the very long term the supply will be mainly SNG. At the research laboratory level it is relatively cheap through scaled down experiments to keep all the options open until the day when they can be closed off.

In order to ensure that the gas is safely transferred to the user it is essential that the transmission and distribution system should be maintained to a high standard. The programme for distribution systems involves constant surveillance for leakage and a programme of replacement. For high pressure transmission systems the need is slightly different; it is necessary to inspect the pipelines from the inside while they are still carrying gas - the On-Line Inspection System. In order to develop effective methods and carry out these inspections a special centre has been established within British Gas.

It is necessary to monitor pipelines for defects which appear after construction and initial testing. Damage from external interference is by far the most numerous of such defects. However there are also the minor defects which initiate and develop in service from growth mechanisms such as fatigue, general corrosion and stress corrosion. It is essential to detect these and take remedial action before any potential problems arise. British Gas is developing an approach that uses the "pig" principle as a basis for a vehicle which carries special equipment for inspection and on-board processing and storage of information for subsequent analysis by computer (101) and then verification by excavation. Several types of detection sensors are in use or under development; magnetic flux to detect defects which are typified by metal loss, for example corrosion, and elastic waves for the detection of crack like defects. The latter development was undertaken on behalf of British Gas by AERE Harwell.

This total programme of surveillance will ensure that the transmission system will be maintained as a valuable asset well beyond the period when the supply has switched over to include SNG.

4.2 SNG from Oil

One method of meeting peak demand for natural gas, or eventually baseload, is to manufacture SNG from whatever feedstocks are available on the open market. It is not an addition to overall energy supplies but the result of converting one fuel into another with inevitably some thermal loss. The loss can be set against the great advantages of a clean, flexible and controllable fuel. We have already seen how steam reforming and hydrogenation had been used in processes for the manufacture of town gas. The feedstocks available in the past were often low boiling distillate fractions and those available in the future might be more difficult to treat.

The catalytic steam reforming of hydrocarbons by the CRG process yields a mixture of gases very rich in methane. The application of this process is limited only by the availability and the cost of suitable feedstocks. Such feedstocks may be relatively expensive in the future but, for peak load applications, that cost may be partially offset by the low capital cost of the necessary plant.

In order to make the process as flexible as possible it was desirable to increase the fraction of crude oil which could be gasified via the CRG route. An important and successful evolutionary development in recent years has been the programme of catalyst and process development to widen the acceptable feedstock range significantly. It has resulted in roughly doubling the potential availability of CRG feedstock from a given crude oil, as shown in Figure 6, with only a slight loss of process efficiency. It is now possible for the first time to gasify kerosene on a commercial scale and the use of gas oil has been demonstrated on a large pilot plant scale. Further developments may extend the range of suitable feedstocks from lighter fractions produced by 'hydrocracking' the heavier fractions of crude oils.

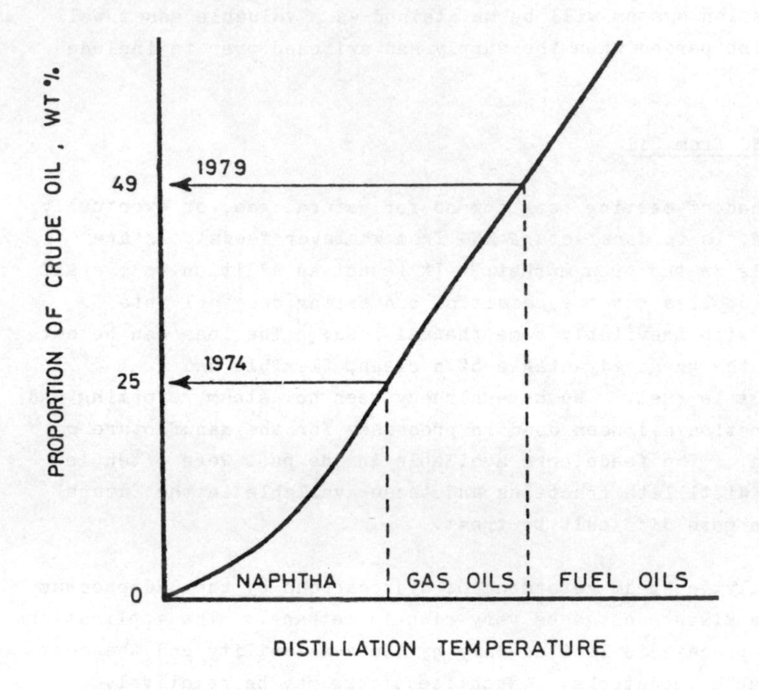

Figure 6 Improvements in feedstock range for the CRG process

The catalytic route to SNG is probably the simplest and most efficient; it will gasify any feedstock that can be volatilised and desulphurised. Other methods are required for residual high boiling feedstocks which may not be amenable to that approach. In the Fluidised Bed Hydrogenator (FBH) shown in Figure 7 a mixture of feedstock and hydrogenating gas is reacted at about 750°C and 70 bar within a recirculating fluidized bed of coke particles. Non-volatile material and carbonaceous residues are deposited on the coke which can be continuously removed from the bed. The FBH process has not yet achieved commercial status but pilot plant and semi-commercial plant experiments with an output of up to 140,000 m^3 per day (1 million ft^3/day) have proved that the system is a feasible process route for gasification of the heavier hydrocarbon feedstocks. Overall thermal efficiencies of up to 75% should be possible.

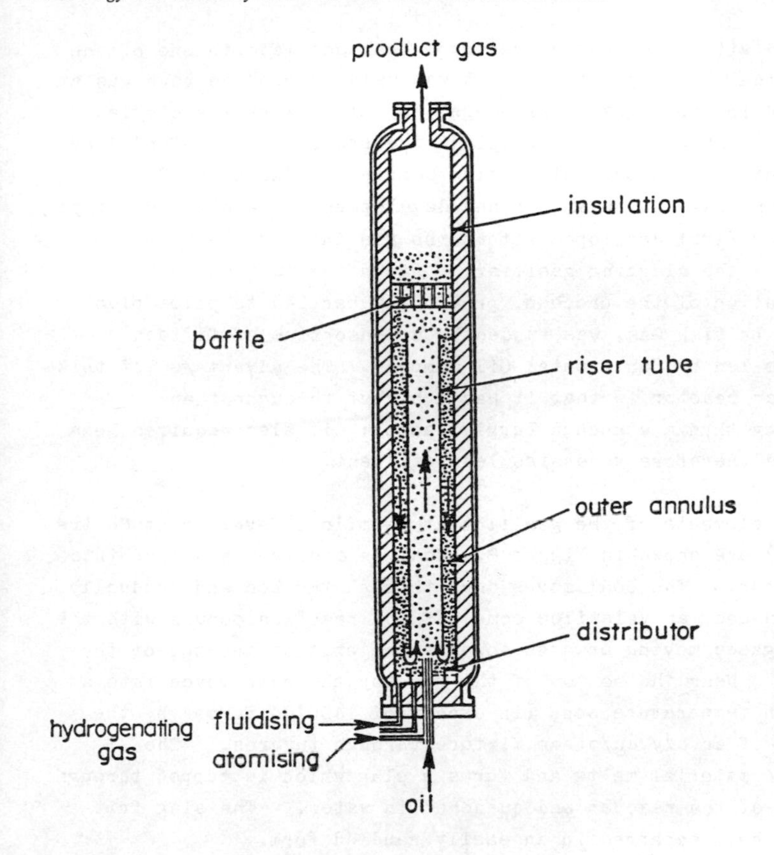

Figure 7 Diagram of the Fluidised Bed Hydrogenator

In the long term, when oil and gas supplies need to be augmented we will need to turn to other sources. In spite of all that is hoped for from what are termed renewable resources the bankers are coal and nuclear energy. Therefore the most likely feedstock for SNG is coal. British Gas has already demonstrated on the commercial scale that SNG can be produced from the gasification of coal and for the first time, anywhere in the world, it has distributed the product in an area of Scotland, in Fife, without the customers in Fife knowing that the changeover had taken place. That experiment, in 1974, was conducted with SNG manufactured from lean gas made in one of the conventional Lurgi gasifiers that was used to produce town gas at Westfield in the 1960s. Although the actual work was carried out by British Gas in its own plant the project was co-funded by a contract with Conoco Methanation Company of Oklahoma USA. Since then there has been further development of a different type of reactor first developed within the gas industry between 1956 and 1964 - the slagging gasifier. Again the full scale demonstration of the process, previously carried to pilot plant scale by British Gas, was funded by a consortium of fifteen companies led by Continental Oil Company. The advantage of this particular reactor is that it has a higher throughput and efficiency than a standard Lurgi reactor; it also requires less steam and therefore generates less effluent.

The main elements of the gasifier at Westfield Development Centre (Plate 4) are shown in Figure 8. Coal is fed from a hopper into the reactor. The coal moves down through the bed and gradually becomes a char as volatiles come off and reaction occurs with the product gases moving upwards towards the exit at the top of the reactor. Near the bottom of the reactor the char moves into a very high temperature zone (in excess of 1400°C) formed by the addition of an oxygen/steam mixture through tuyeres. The inorganic material melts and forms a slag which is tapped through the base of the reactor and quenched in water. The slag frit can then be discharged in an easily handled form.

During the sponsored project the reactor completed a three year test programme including a continuous run of twenty-three days. That was quite a long period for such an advanced plant and since that time British Gas has continued its own development programme on the design and operation of the gasifier. Such a programme requires a full range of support work on the fundamental chemical reactions involved, the development of suitable materials for all aspects of the reactor performance, characterisation of the coals, ashes, slag frit, the gas formed and the effluent which might require treatment. In fact a study of the effluent from this slagging gasifier was the subject of an International Energy Agency project which was completely planned and executed by British Gas.

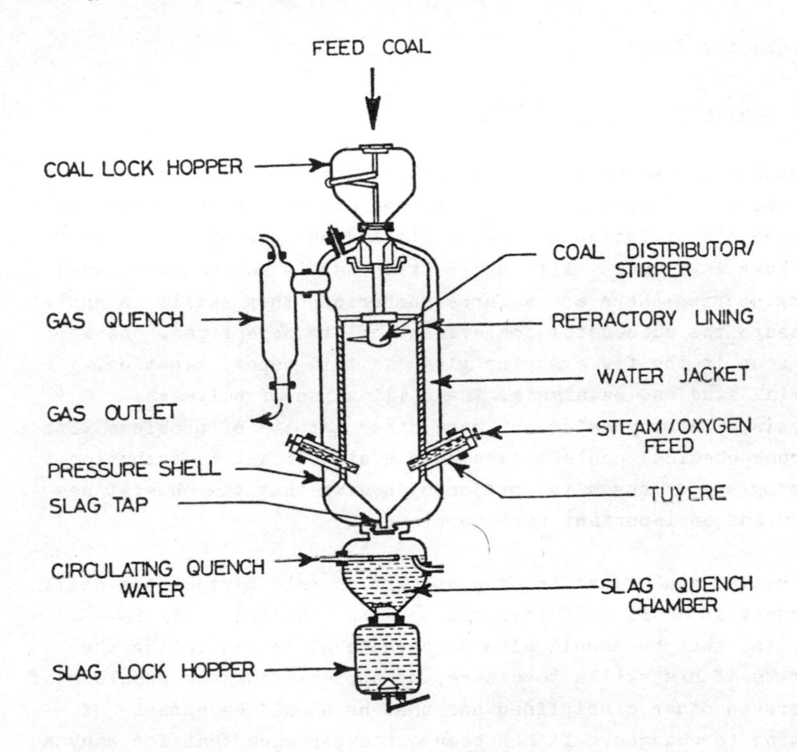

Figure 8 Diagram of the Slagging Gasifier

Another contract, funded this time by the Electric Power Research Institute of the USA, was carried out to assess whether or not the slagging gasifier could respond adequately to load changes. The tests were successful and proved the reactor could be used effectively for combined heat and power generation.

These studies on the total gasification of coal, the effluents formed, the clean up methods for the gas, process routes for methanation of the lean gas, together with the CRG and FBH routes described above, give British Gas several options to meet short and long term demands for energy in the form of SNG. The need for SNG is dependent on many factors such as energy use, conservation and new finds of natural gas but being prepared ensures that the industry has a range of choices when it comes to planning the future.

5. Challenge for the Chemist

The enormous changes which took place within the gas industry over the past twenty years, and indeed are continuing today, have required the evolution of new skills and changes of direction for all those involved. Although British Gas is an industry based on engineering there are a large number of other skills essential to ensure the successful achievement of its objectives. As we have seen in the few examples given in this paper, kinetics, material studies, catalysis, the utilisation of polymers, analysis, thermodynamics and many other aspects of problems with a strong chemical content have made a significant contribution to the progress of essential projects showing that the chemist has always had an important part to play.

The initial qualification of a chemist is only part of the skills necessary if he is to fulfil his role in industry. It is essential that he should also be capable of communicating the benefits of his skills to others, of appreciating the problems of workers in other disciplines and that he should be capable of adapting to changes. It has been our experience that for many a chemist the excellence of his training and his willingness to adapt have led to new opportunities and successes in completely new fields.

As British Gas looks forward to the next twenty years, and ultimately to a time when we will have come full circle again to manufacture gas from coal, there are still many challenging problems to be solved. If the transition is to be as successful as those in the past then well-trained, inquisitive, progressive and adaptable chemists will be required to play their part, with colleagues of other disciplines, in continuing the evolution and revolution within the gas industry.

Chemistry in the Development of Nuclear Power

By J. A. C. Marples, R. L. Nelson, P. E. Potter,
and L. E. J. Roberts, C.B.E.
AERE, HARWELL, DIDCOT, OXFORDSHIRE OXII ORA

The fission reaction of uranium-235 bombarded with neutrons was a discovery of two radiochemists and chemists and chemical engineers have played a vital role in the practical exploitation of this new dimension of nuclear physics. "Chemists, chemists and still more chemists" was the wry comment of a senior physicist in the early days of the industry. The excitements of the early days have been well documented and have passed into history - the characterisation of a whole new series of elements, the actinides, of the chemistry and radiochemistry of the fission products, and the development of practical plants for plutonium separation on the basis of experiments carried out with a few milligrams of material. Large scale reprocessing techniques are still vital to the economy of nuclear power, and the work done for the nuclear industry has contributed much to the general science and technology of ion exchange, solvent extraction and the separation of isotopes. But these subjects will be dealt with by Dr. Wilkinson in the following lecture.

The part that chemists have played and still play by no means ends there; chemists are found in every reactor development project. The characteristic requirements of nuclear technology, in reactors or in the associated chemical plant, are those of high reliability, long life and ease of maintenance. All this must be achieved in the unique circumstances of high radiation fields, and a characteristic dimension in all materials work connected with nuclear power is the study of susceptibility or resistance to radiation damage. Thus a great effort has gone to understanding and, now, control of the complex sequence of reactions of the radiolysis products of CO_2 with graphite, reactions which take place within the pores of the graphite itself and which could determine the safe life of the gas-cooled reactors we are now building. An even larger effort has been devoted, internationally, to the characterisation and control of the complex corrosion reactions in water-moderated reactors, again under the irradiation conditions which will be met in practice. This type of work, with the stringent standards of purity demanded by nuclear considerations, has led to the need for very advanced analytical techniques, and it is no accident that one of the largest contributions which Harwell has made to chemistry is in the analytical area. The heterogeneous nature of the topics just mentioned has led to the very detailed study of surface chemistry and surface reactions becoming another characteristic subject which we have sought to support by steady investment in the many sophisticated physical techniques which have been developed.

As the technology has advanced, more emphasis has been placed on environmental control. This requires, for instance, better knowledge of the chemistry underlying the complex phase changes which occur in fissile material during irradiation; it requires the development of technologies in fuel processing which minimise radiation dose to operators and it requires careful attention to all aspects of radioactive waste management and disposal. I want to illustrate the type of contribution which chemistry is now making to these vital parts of the nuclear power programme by talking for a short time of three examples of recent work for, indeed, to try to cover the whole field would result in a catalogue but not a lecture.

1. Phase Equilibria and Thermodynamic Properties of Oxide Fuels

One subject in which chemists have been and are still active is the study of the many complex changes of chemical consitution which occur during the irradiation of an oxide nuclear fuel in both thermal and fast-breeder nuclear reactors. An increasing knowledge of the thermodynamic properties of individual systems is contributing to an understanding of these phenomena.

The fuels are (1) uranium dioxide for thermal reactors, e.g. light water reactors (Pressurised Water Reactors (PWRs) and Boiling

Water Reactors (BWRs)) and gas-cooled reactors (Advanced Gas-cooled Reactors (AGRs)), and (2) solid solutions of uranium dioxide and plutonium dioxide for the sodium-cooled fast breeder reactor. A typical LWR fuel consists of dense pellets of uranium dioxide contained in a Zircaloy can, whilst the AGR fuel consists of uranium dioxide pellets contained in cans of stainless steel. These thermal reactor fuels are enriched in the uranium-235 isotope. For the existing fast reactors the fuels are pellets of a single phase solid solution of uranium and plutonium dioxide.

A possible development for fast reactor fuels which is discussed later in this paper will be the use of gel-precipitated spheroids of the solid solutions packed into stainless steel cans, instead of pellets of fuel. Depleted urania is used in the fabrication of the breeder pins in a fast reactor core.

An irradiated fuel is one of the most complicated high temperature systems found in industry today. The transformation of the original actinide elements into fission product elements in the presence of steep temperature gradients leads to mass transport and chemical interactions, the understanding of which presents a formidable but fascinating challenge to the chemical thermodynamicist and is essential to the successful exploitation of a nuclear fuel. Many of the chemical effects, such as the nucleation of new phases, are more pronounced in fast reactor fuel pins than in those of a thermal reactor, because in the former the fuel centre temperatures are higher, the temperature gradient is steeper, and the proportion of actinide atoms which undergo fission is higher. The temperature of a fuel pin will be 2000-2400°C in the central region and 700-800°C at the fuel periphery. The proportion of actinide atoms which undergo fission in a fast reactor is ca. 10%, and in thermal reactor systems is ca. 2-4%. Although the number of displacements of each actinide atom in the lattice of the fuel is ca. 10^{2} per 1% burn-up, the damage to the lattice is essentially annealed out at the temperatures of fuel operation.

Over thirty elements are produced during fission and their chemical forms can depend upon the initial composition of the fuel. The important effects to consider are those which determine the compatibility between the fuel and its clad and consequently determine the appropriate level of burn-up of the fissile atoms within a fuel pin.

1.1 The Actinide Oxides. The binary oxide systems of the actinide elements are extremely complex because, in addition to the large number of phases in these systems, many of them exist over a wide range of composition. The actinide oxides of the greatest technological significance are the dioxides (MO_2); these oxides crystallise with the fluorite (CaF_2) structure. The dioxides with this structure are known for the elements from thorium to californium. With the exception of PaO_2 all the actinide dioxides can lose oxygen at high temperatures and become markedly hypostoichiometric (MO_{2-x}), with oxygen vacancies in the anionic lattice. Protactinium and uranium dioxides also take up oxygen to form hyperstoichiometric oxides (MO_{2+x}) with the additional oxygen in interstitial positions in the fluorite lattice. The positions of the interstitial oxygen atoms in uranium dioxide[1] were determined by neutron diffraction.

1.2 The Uranium-Oxygen System. The thermodynamic features of the uranium-oxygen system have been critically examined recently[2] and here discussion is confined to the region of the system for uranium dioxide, the composition range of which is strongly dependent on temperature. Phase diagrams for the hypo- and hyper-stoichiometric regions of uranium dioxide are shown in Figure 1. The difference in energies of the vacancy and interstitial defects in hypo- and hyper-stoichiometric uranium dioxide is reflected in the large change in partial molal Gibbs energy of oxygen ($\bar{G}_{O_2}$) or oxygen potential ($\bar{G}_{O_2} = RT\ell nP_{O_2}$: R is the gas constant, T the temperature in K, and P_{O_2} the oxygen pressure in atmospheres) in traversing the stoichiometric composition. Both types of vacancies[3] are most probably doubly charged[4]. Electronic defects must also be produced on oxidation and reduction of $UO_{2.00}$[4]. Electrons in the conduction band are created in reduced UO_{2-x} whilst holes are present in the valence band of UO_{2+x}. Both the valence and the conduction bands in uranium dioxide have cationic character, and the highly polar nature of the oxide favours the localisation of electronic species in individual cation sites - the formation of small polarons. The holes present in UO_{2+x} correspond almost certainly to distinct U^{V} cations and the electrons in UO_{2-x} to U^{III} cations.

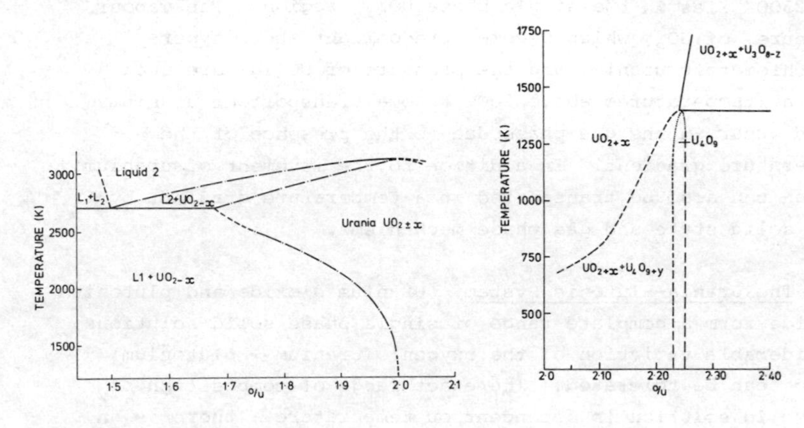

Phase Diagram of Hypostoichiometric Urania

Phase Diagram of Hyperstoichiometric Urania

Figure 1

The oxygen potential is an important parameter in determining the compatibility between the oxide fuel or breeder and the cladding of the pins. The oxygen/metal ratio of the fabricated uranium dioxide is as close as possible to 2.00 but, as will be noted below, the oxygen:cation ratio can increase with increasing fission of the uranium and plutonium atoms.

In order to understand the behaviour of oxide fuels, a knowledge of their thermodynamic properties is required up to and beyond the melting points; that of UO_2 is 3120K. At these high temperatures the vaporisation behaviour has to be considered. Although the vapour from $UO_{2.00}$ is mainly UO_2(g), mass spectrometric examination of the gas phase has shown that there are small quantities of UO(g) and UO_3(g) present, as well as much smaller amounts of U(g), O(g) and O_2(g). As uranium dioxide becomes more hypostoichiometric the proportion of UO(g) increases and conversely, as it becomes more hyperstoichiometric the proportion of UO_3(g) increases, and the pressure of UO_3(g) eventually become greater than that of UO_2(g). There will therefore be a congruently vaporising composition (CVC) at which the compositions of the gas and solid are identical and the total gas pressure is a minumum. The CVC varies with temperature and up to ca. 2500K lies in the single phase UO_{2-x} region. The vapour pressure of UO_3, which becomes predominant above hyper-stoichiometric urania, and the pressure of UO_2(g) are such that at temperatures above 2000 K some transport of uranium could occur in the gas phase due to the presence of the temperature gradient. In addition to the movement of uranium, oxygen can also be transported in a temperature gradient by both solid state and gas phase mechanisms.

1.3 The Urania-Plutonia System. Uranium dioxide and plutonium dioxide form a complete range of single phase solid solutions; considerable variation of the oxygen: (uranium + plutonium) ratio can be tolerated. The exact range of composition of the solid solution is dependent on temperature[5]; there is an appreciable area of single phase fluorite-structured solid solution which occurs in solutions cooled to room temperature. The solid solution can be both hypo- and hyper-stoichiometric with respect to oxygen; in the hypostoichiometric region Pu^{III} cations (and at high temperatures some U^{III} cations) are present, whilst in the hyperstoichiometric region U^{IV} cations are oxidised to U^{V} cations.

The compositions of the solid solution which are of interest to the reactor technologist are those with Pu: (U + Pu) ratios of 0.15 to 0.40 for fast reactor fuels and for the fast reactor breeder and thermal reactor fuels those with Pu: (U + Pu) ratios of up to ca. 0.1. The O: (U + Pu) ratio of fast reactor fuels is usually 1.98-2.00. In addition to the knowledge that the fluorite solid solution will remain single phase over the range of compositions and temperatures likely to be encountered within the reactor core, a knowledge of the variation of thermo-dynamic oxygen potential with O: (U +Pu) ratic and plutonium concentration is required as for the uranium dioxide phase in order to assess the likely behaviour of the fuel during operation of a reactor. Some experimental measurements,[6] which suggested that the oxygen potentials of the hypostoichiometric oxides were dependent only upon the average valency of the plutonium ions, have been substantiated by recent measurements[7] using a thermogravimetric method in the temperature range 1273-1473K on hypostoichiometric solid solutions with Pu: (U + Pu) ratios of 0.1, 0.25 and 0.40. Some of these recent data are shown in Figure 2. Such a relationship makes possible the extrapolation of oxygen potentials up to the temperatures encountered in fast reactor fuel pins. Further measurements of oxygen potentials at temperatures above 2000K for a range of plutonium concentrations are however required on hypostoichiometric solid solutions; the existing measurements[8] at 2050K for plutonium concentrations corresponding to Pu: (U + Pu) ratios of 0.15, 0.20, 0.25 and 0.30 and with Pu valencies of between ca. 3.5 and 3.8 indicate that, at this temperature, there may be an influence of Pu concentration as well as of average valency of Pu cations on the oxygen potential. The recent measurements of oxygen potential for hyperstoichiometric oxides[9] indicate that the oxygen potential of these solid solutions depends not only on the valency of the uranium cations but also on the uranium concentration.

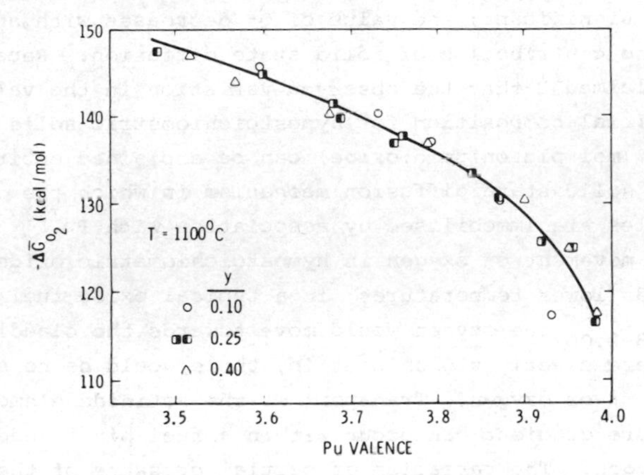

Variation in the Oxygen Potential of $U_{1-y}Pu_yO_{2-x}$ with Pu Valence at 1100°C

Figure 2

The possible redistribution of uranium by gas phase transport in a pin with a high centre temperature has been mentioned; the redistribution of oxygen and of uranium and plutonium in the mixed oxide solutions must be considered. Redistribution can result in marked changes of chemical constitution. Redistribution of oxygen can occur in UO_2 and $(UPu)O_2$ solid solutionsby both gaseous and solid state mechanisms[10]. A common form of representation of the distribution of oxygen in a temperature gradient is the equation

$$\ell n\ x = \frac{-Q^*}{RT} + \text{constant}$$

where x is the deviation from stoichiometry, R is the gas constant and T is the temperature (K). The energy parameter Q* is frequently called the 'heat of transport'. A possible mechanism for the transfer of oxygen within an oxide fuel is through the gas phase by mixtures of CO/CO_2 and H_2/H_2O. The formation of these gases would be due to the presence of carbon and hydrogen impurities within the fuel material. A constant CO/CO_2 or H_2/H_2O ratio would be maintained by thermal convection through interconnected porosity in the fuel, and the fuel at every temperature would be in equilibrium with the gaseous mixture. A more general approach has been developed in which solid state transport of oxygen is also considered and indeed as the oxygen:metal ratio of the solid solution decreases the latter mechanism becomes more and more significant; the value of Q* decreases with an increase in the contribution of solid state diffusion. Recently it has been claimed[11] that the observed variation in the value of Q* with initial composition for hypostoichiometric solid solutions (25% mol plutonium dioxide) can be explained entirely in terms of a solid state diffusion mechanism in which the oxygen vacancies are immobilised by association with Pu^{III} cations. The movement of oxygen in hypostoichiometric oxides will be towards lower temperatures; in a typical oxide fuel $U_{0.7}Pu_{0.3}O_{1.98-2.00}$ the oxygen would move towards the cladding. If the fuel were exactly stoichiometric, there would be no significant transport of oxygen. Transport of the actinide elements in a temperature gradient can occur within a fuel pin in addition to that of oxygen. The variation of partial pressure of the individual gas phase species with stoichiometry above the solid solution at 2000K is shown in Figure 3. It will be seen that the gas phase is rich in Pu at oxygen:metal ratios less than ca. 1.97, and rich in U for oxygen:metal ratios greater than ca. 1.97. Thus in irradiated fuels with oxygen:metal ratios between 1.98 and 2.00 the hot centre of the fuel is enriched in plutonium due to the preferential loss of UO_2(g) and UO_3(g). The phase equilibria and thermodynamic data for the oxides are an essential starting point for the understanding of the changes which occur during irradiation.

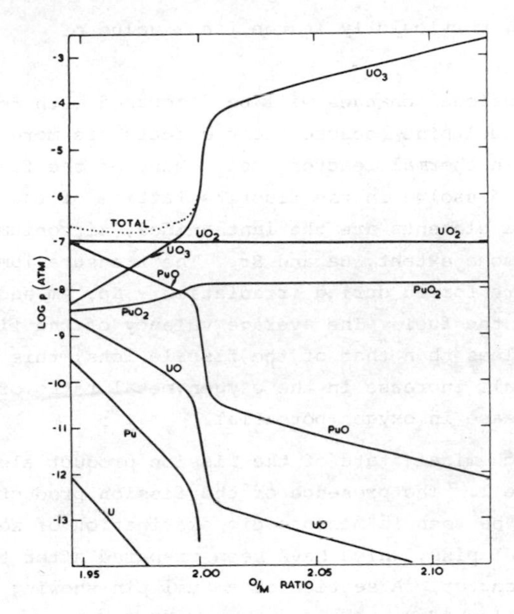

The Variation of Pressure with Composition of the various gaseous species above $U_{0.85}Pu_{0.15}O_{2\pm x}$ at 2000K

Figure 3

1.4 <u>Changes in Chemical Constitution During Irradiation</u>. As more and more fissile atoms are burnt, compositional changes become important. A simple approach which allows a prediction of the chemical state of irradiated oxide fuels is to take the total likely cations of the fission product elements and combine them with the available oxygen atoms in the order of the thermodynamic stability of these oxides. The requirement of the calculation with the fission product element oxides is that the total value of the partial molal Gibbs energy of oxygen or thermodynamic oxygen potential ($\bar{G}_{O_2}$) should be a minimum. In practice, such a condition would only hold for a small domain of the system; the global condition for equilibrium in an isothermal system would be perturbed by the temperature gradient.

It is important in any assessment of changes in chemical constitution that compounds containing more than two components are not neglected, otherwise considerable inaccuracies could be present, particularly in the final value of $\bar{G}_{O_2}$.

The constitutional changes will be discussed with respect to fast reactor fuel pins because their effects are more pronounced than in thermal reactor fuel. Many of the fission product elements dissolve in the fluorite lattice of the fuel; among those elements are the lanthanides, zirconium, niobium and, to some extent, Ba and Sr. The transuranium elements which are formed during irradiation - Np, Am and Cm - also dissolve in the fuel. The average valency of the fission product ions is less than that of the fissile ions; this leads to an overall increase in the oxygen:metal ratio of the fuel and an increase in oxygen potential.

The likely chemical state of the fission product elements is shown in Table 1. The presence of the fission product phases can often be seen in microscopic examination of sections of irradiated fuel pins, which have been prepared after their removal from a reactor. A section of a fuel pin showing the presence of an alloy of Mo, Tc, Ru, Rh and Pd is shown in Plate 1. The composition of this 5-component alloy can vary somewhat because in an environment of relatively high oxygen potential some Mo can oxidise to MoO_2. This reaction was considered as being responsible for the buffering of the oxygen potential within the fuel, but it is possible that reactions such as the formation of caesium uranate and those between fission products, fuel and stainless steel could buffer the oxygen potential[12]. The volatile fission product elements such as the alkali metals, the halogens and Te migrate to the colder regions of the fuel pins and are the elements most frequently found in the fuel-clad gap.

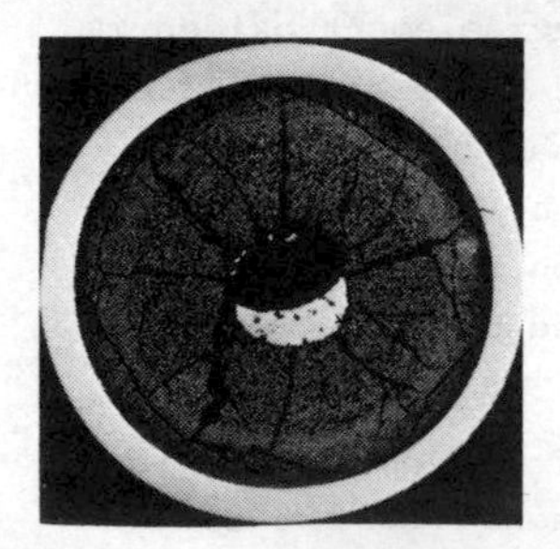

Plate 1

Post Irradiation Examination of a Fuel showing Metallic Inclusions of the Mo-Tc-Ru-Rh-Pd alloy in the central void

Table 1

The Possible Chemical State of the Fission Product Elements in Oxide Fuel

Fission Product	Chemical State
Kr, Xe	Remain in elemental form
Br, I	RbI, RbBr, CsI, CsBr
Rb, Cs	As halides and Cs_2MoO_4, $Cs_{2-z}UO_{4-x}$ $(Cs,Rb)_y$ $(Cr,Mo,U,Pu)O_4$ $(y \simeq 3)$
Se, Te	Rb_2Te, Rb_2Se, Cs_2Te, Cs_2Se
Sr, Ba	SrO, BaO and $Ba_{1-x}Sr_x$ $(Zr_{1-y-z}U_yPu_z)O_3$
Sr, Nb	Dissolved in $U_{1-y}Pu_yO_{2-x}$ (fuel matrix)
Y, La, Ce, Pr, Nd, Pm, Sm, Eu, Gd, & actinides Np, Am, Cm	Dissolved in fuel matrix
Mo, Tc, Ru, Rh, Pd	Mostly an alloy (single phase): some Mo can be as MoO_2 or in complex phases (see Rb, Ca); some Pd can be combined with Te
Ag, Cd, In, Sn, Sb	Fission yields low. Ag found with Pd

Although the unirradiated fast reactor fuel is compatible with stainless steel, the presence of a liquid phase of fission products containing, probably, at least Cs, Mo, Te, I and O gives a means for reaction with the protective layer on the steel. In fact, two types of clad corrosion are found in irradiated oxide fuels. One, in which the whole surface of the steel is attacked, is called variously 'matrix', 'broad front' or 'surface ablation' since the maximum penetration so far noted is about 10% of the clad thickness. The other and more serious form of corrosion is 'intergranular attack' which can penetrate to 25% of the clad thickness. Indeed, in very active cases this, or a related type of attack, has led, in very highly rated pins with vibrocompacted fuel, to the opening of a crack right through the can. Where this occurs high, but variable, local concentrations of Cs, Mo, Cr, Te, I and O are observed in grain boundaries, with chromium and molybdenum depletion of steel. The nature of the products formed and the reaction mechanism are still under investigation. It seems probable that a phase

$$(Cs,Rb)_y (Cr, Mo, U, Pu)O_4$$

with y ≈ 3 may be involved, perhaps as a solid, perhaps in solution as a caesium-rich 'soup'. The deeper intergranular attack may result from the penetration and reaction of the Cs-rich liquid and perhaps of a gaseous species aided by stress corrosion and the presence of $Cr_{23}C_6$-type precipitates in the sensitized steel. The deeper attack is usually associated with Te:Cs ratios higher than that corresponding to the fission yields of these elements, but whether this is a cause or effect is not clear.

Our knowledge of the actinide oxide and fission product chemistry thus allows the prediction with some certainty of the new phases which are likely to appear during irradiation of an oxide fuel under different conditions. There is still much fascinating chemistry to do. The understanding of the conditions under which deleterious reactions can occur will lead to better control and to an optimised regime for fast reactor fuel management.

2. Gel Precipitation

We now turn from fuel behaviour to the preparation and fabrication of oxide fuels. Until the present time a mixed oxide $(U,Pu)O_2$ has been universally prepared as pellets by a conventional ceramic processing route[13], which involves the preparation of fine powders of UO_2 and PuO_2, their blending, addition of binder, granulation, pressing to pellets, removal of binder by debonding and final sintering. Although this processing sequence is operable under closely controlled laboratory conditions, there are penalties in handling finely divided powders because of (a) the need to restrict radiation dose to operators and maintenance staff, (b) the need to recycle material through the process, and (c) the need to move to remote operation and maintenance concepts in future, larger scale plants.

For these reasons there has been a strong incentive in a number of countries to develop a 'wet' process which eliminates the use of fine powders. By ensuring that the process vessels themselves form the 'primary containment', this method has potential for reducing the need to recycle material and for the provision of a remotely operable plant[14].

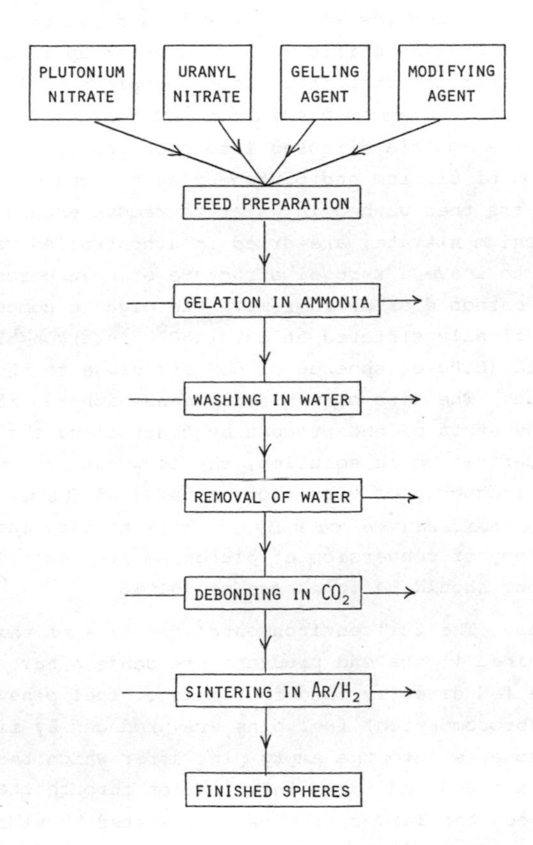

Outline Chemical Flow Diagram for Gel Precipitation Route to $(U,Pu)O_2$
Figure 4

2.1 The Gel Precipitation Process. The 'wet' route selected by the UKAEA and BNFL for development is gel precipitation[15] and an outline chemical flow diagram is shown in Figure 4. Concentrated solutions of uranyl (U^{VI}) nitrate and acidic plutonium (Pu^{IV}) nitrate are obtained directly from the reprocessing plant and without any further treatment are mixed with aqueous solutions of an organic polymer (the 'gelling agent') and another organic compound (the 'modifying agent') which is capable of complexing the heavy metal ions and preventing premature reaction between them and the organic polymer. The composition of this mixed solution is selected to give appropriate rheological properties so that the liquid can be pumped through a vibrating orifice so as to break up into uniformly sized spherical droplets. These droplets fall into a column of concentrated ammonia solution and solidification to a gel occurs as the ammonia diffuses into each sphere causing the precipitation of uranium and plutonium as hydrous oxides. The gel spheres are then washed in water to remove excess ammonia and ammonium nitrate, are dried in a controlled manner to remove water to leave a 'xerogel' structure of open porosity, are debonded in carbon dioxide to remove the organic compounds present and are finally sintered at ca. 1450°C in argon/5% hydrogen to yield $(U,Pu)O_2$ spheres of density close to the theoretical value. The size of the fully dense spheres is determined at the start of the process by controlling the heavy metal concentration in solution, the flow rate through the orifice and the frequency of vibration; spheres of 800µm and 80 µm final diameter have been made routinely. . Work to date indicates that the efficiency of conversion of plutonium from solution into final product should be in excess of 99.9%.

2.2 Applications. The full environmental benefits of this process are obtained if the end products are dense spheres and these can be fed directly into fast reactor fuel pins. 'Vipak' (i.e. vibrocompacted) fuel pins are produced by first loading 800 µm spheres into the empty pins after which the 80 µm spheres are added and their infiltration through the interstices between the larger spheres is assisted by vibration; the loading obtained by using two sizes of spheres whose diameters differ by a factor of 10 is ca. 80% of theoretical. This fuel is one of the candidates for the Commercial Fast Reactor. Alternatively, spheres prepared by gel precipitation may be converted into pellets, without compromising the environmental advantages of the 'wet' route.

Since the gel precipitation process appears capable of being operated continuously through a series of columns with remote control it is more likely than the conventional powder process to be able to handle more active grades of plutonium such as come from light water reactors, fast breeder reactors or as a result of deliberate retention of fission products, to increase the proliferation resistance of the route.

The gel precipitation process is also being considered as the finishing process for the new THORP plant at Windscale, partly because of the advantages of the process and partly because of the excellent handling and storage characteristics of spherical material.

The use of a mixed uranium-plutonium solution as feed to the gel precipitation process is entirely consistent with a co-conversion policy in which these two elements are processed together throughout the fuel cycle.

2.3 Detailed Chemical and Physical Changes During the Gel Precipitation Process. Although the desired fuel for the fast reactor is $(U,Pu)O_2$, studies have also been conducted on the preparation of UO_2 and $(U,Th)O_2$ in order to yield information on mechanism and structure.

2.3.1 Precipitation, Washing and Drying. When a droplet of feed solution makes contact with the concentrated ammonia solution, precipitation of uranium and plutonium as hydrous oxides starts at the surface and the reaction interface moves inwards as ammonia diffuses through the already-precipitated gel. The ammonia first neutralises the excess acidity present in the feed as a result of using an acidic solution of plutonium nitrate and then supplies the hydroxyl ions necessary for the precipitation. In the absence of the organic polymer a shapeless, gelatinous precipitate is formed whereas the gelling agent preserves the spherical shape and leads to the production of a tough gel particle which can readily withstand handling during the subsequent steps of processing. During the gelation stage the organic polymer has become insoluble in water and this is taken as evidence for a strong interaction between the hydrous oxides and the polymer. During the washing process the ammonium nitrate product is removed from the spheres along with the modifying agent and excess ammonia. The drying process then removes the bulk of the residual water and, if this is carried out in a controlled manner, open pores can be maintained that permit the further processes of debonding and sintering without the build-up of internal pressure within the spheres as a result of gas generation.

The chemical and physical structure of xerogel (i.e. dry gel) spheres will now be considered in detail.

2.3.2 <u>X-ray Diffraction Studies</u>. If the gels have been in contact with the ammonia for precipitation and water for removal of soluble products for minimum periods the X-ray diffraction pattern only contains broad peaks (see Figure 5) which relate to crystallites of size ca. 20 Å. However, if the wet gel spheres are allowed to age in ammonia or water the peaks sharpen up (Figure 5) as crystal growth occurs. In mixed oxide precipitates no pattern is seen for the thorium or plutonium component and only the diffraction pattern for 'ammonium diuranate' is seen. Detailed chemical analyses indicate that the composition of ADU in the wet gel spheres as precipitated is $UO_3.xNH_3.(2-x)H_2O$ with x approaching 0.6 though after washing in water the value of x decreases to 0.3. It thus appears that the uranium is precipitated in the crystalline ADU phase but both thorium and plutonium form amorphous hydrous oxides $MO_2.xH_2O$ possibly closely connected with the immobilised polymer molecules.

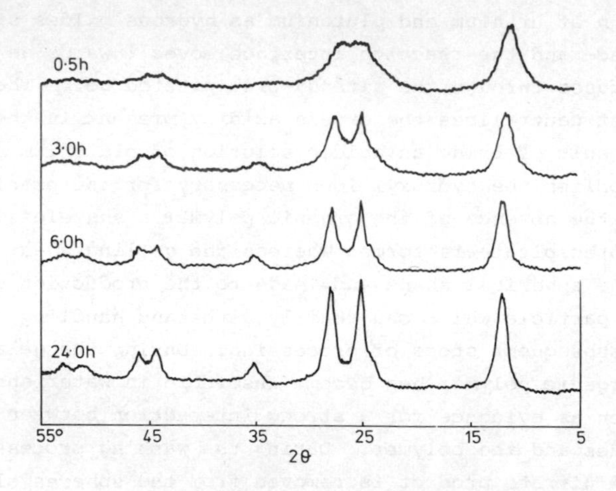

X-Ray Diffraction Patterns Showing the Effect of Ageing Time in Ammonia for ADU Gel-Precipitated Spheres

Figure 5

The role of the polymer can also be partly understood from the X-ray diffraction patterns which compare precipitates formed in the absence of polymer with those of the gel precipitation process described above. In the absence of polymer the ADU crystallite size is typically 150 Å as precipitated and is not affected by ageing, whereas the presence of polymer clearly reduces the mean crystallite size initially, suggesting a chemical interaction between the polymer and the surface of the crystallites. One might expect this behaviour in view of the use of such water soluble polymers as flocculating agents in water treatment. A further point concerns the effect of the tetravalent ions, thorium-IV or plutonium-IV, on the size of the ADU crystallites (see Figure 6). As the concentration of the tetravalent ion in the feed increases, the ADU crystallite size decreases. The tetravalent ions interact more strongly with the polymer (in a cross-linking process) than the uranyl ions and it is possible that the polymer-tetravalent ion complex has a more inhibiting effect on ADU crystal growth than polymer alone. The mixed oxide gel spheres are mechanically much tougher than urania gel spheres at the wet stage and this is a reflection of the polymer-tetravalent ion interaction. Evidence for a strong polymer-tetravalent ion interaction comes partly from rheological studies of the feed solution and partly from the dissolution characteristics of dry gel. In this latter case ADU gel dissolves completely in nitric acid whereas with the (U,Th) and (U,Pu) gels nitric acid dissolves the ADU component completely but the polymer, along with a fraction of the thorium and plutonium, remains insoluble.

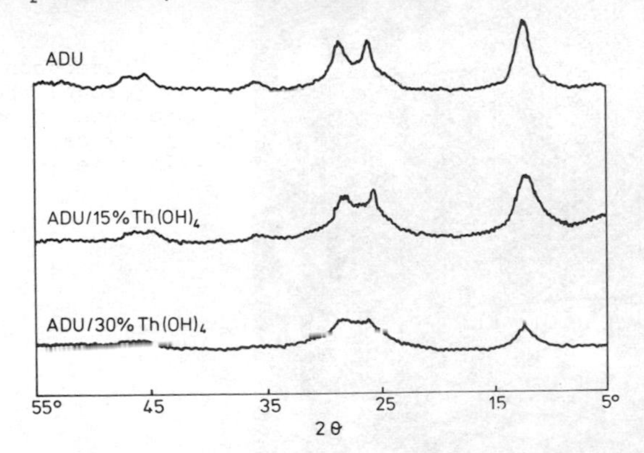

X-Ray Diffraction Patterns Showing the Effect of Tetravalent Ions on Crystal Growth in Gel-Precipitated Spheres (aged 3 hours in ammonia)

Figure 6

2.3.3 Scanning Electron Microscopy. After drying the wet gel spheres in a controlled manner to produce the xerogel, a hard, rigid structure is formed. This xerogel is still highly porous and gas adsorption studies have indicated a specific surface area of ca. 100 m^2g^{-1}.

SEM studies of xerogel fracture surfaces indicate that the spheres are made up of agglomerates of ca. 1500 Å in diameter (Plate 2) and each agglomerate contains grains about 100-200 Å in size. This grain size is larger than the size of the individual ADU crystallites, as determined in the unaged gel, but after ageing the crystallites are of comparable dimensions to the visible grains. Pure thoria produced by gel precipitation, in which the hydrous thoria is amorphous, still possesses 100-200 Å grains and this suggests that in the mixed oxide each grain represents the 'basic unit' of precipitation within which ADU, hydrous Th or Pu oxide and polymer are immobilised as the ammonia diffuses through the droplets in the precipitation column. The polymer molecule inhibits growth of ADU but, during ageing, crystal growth does occur and is subsequently limited by the grain size. The individual grains are probably held together within the aggregates by polymer-polymer interactions.

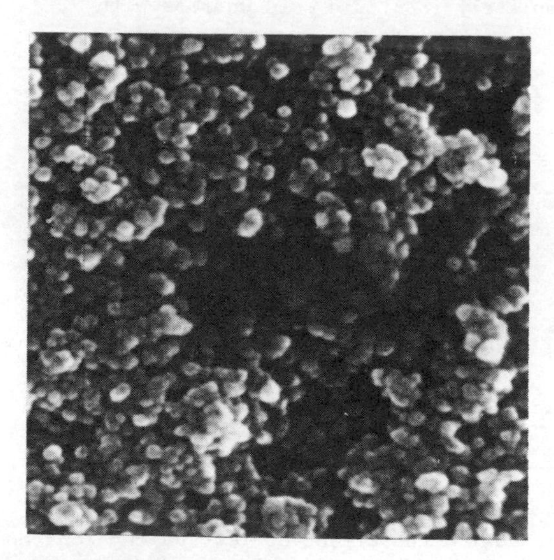

Plate 2

Stereoscan of ADU Xerogel showing the presence of aggregates of ca. 1500 Å in size (x 20,000)

2.4 Changes During Debonding and Sintering. The debonding process involves heating the xerogel spheres to a temperature of ca. 800°C in an atmosphere of carbon dioxide, primarily to remove the organic polymer from the oxide structure.

2.4.1 X-ray Diffraction. Many of the changes in the chemical structure can be deduced from a combination of elemental analysis and the X-ray diffraction patterns obtained after debonding to different temperatures. The main change up to 250°C is loss of water and ammonia with partial conversion of ADU to UO_3. Above 200°C the loss of organic material becomes significant, the UO_3 is reduced to $UO_{2.25}$ at about 300°C and the structure passes through an amorphous stage. By 400°C a crystalline $UO_{2.25}$ structure is established with a similar crystallite size to the earlier ADU structure, presumably because the crystallite size is physically limited by the size of the grains (i.e. 100-200 Å). Above 500°C the skeletal carbon remaining from the decomposition of the polymer is removed by oxidation with carbon dioxide and $UO_{2.25}$ is reduced to UO_{2+x} where x tends to 0.05. Incorporation of Th or Pu into the UO_{2+x} structure to yield the mixed oxide occurs between 500 and 800°C as shown by changes in the lattice constant.

2.4.2 Scanning Electron Microscopy. During the debonding process the grain size does not show any increase up to 650°C but the crystallites within the grains have grown. The grains are by then fully dense (ca. 11.1 g cm^{-3} for $(U,30\%Pu)O_2$) but there has been no sintering between grains. Above 650°C sintering between grains starts and the crystallite size increases dramatically.

After debonding the spheres are sintered in argon/5% hydrogen at 1450°C and this process removes residual carbon to < 50 ppm, adjusts the O/M ratio to ca. 1.98 for $(U,Pu)O_2$ and completes the densification of the entire structure to yield bulk densities of 98 ± 2% of the theoretical value.

These studies of the physical and chemical changes during the different stages of the gel precipitation process give us greater control over continuous operations and they have indicated how specific properties such as density, pore volume and crystallite size, can be modified, particularly at the intermediate stages. This process has potential for applications outside the nuclear industry; three such areas are the use of gel carriers for pharmaceuticals, porous ceramic substrates for catalysts and adsorbents.

3. Glasses for the Solidification of High Level Waste

When nuclear fuel elements have come to the end of their useful life, they are removed from the reactor, stored for a few months to allow the worst of the radioactivity to decay, and then reprocessed to recover the 'unburnt' uranium and the plutonium, which is a useful nuclear fuel.

This very efficient process leaves 99% of the fission products in the aqueous phase which becomes the Highly Active Waste (HAW). This latter also contains residual amounts of plutonium and uranium (about 0.1% and 0.002% of the amounts originally present), and also most of the other actinide elements that have been generated in the reactor (neptunium, americium and curium). Most of the HAW in the UK has come from the reprocessing of fuel from the Magnox reactors. In addition to the active isotopes mentioned above, this also contains relatively large amounts of magnesium (from pieces of can that are not removed and which get into the dissolver), aluminium (from grain refining additions to the fuel), and iron (from corrosion of the plant). A typical waste composition (expressed in wt% oxides) is:

FPO_x	MgO	Al_2O_3	Fe_2O_3	Cr_2O_3	NiO	ZnO	UO_{2+x}
39.1	24.8	20.0	10.6	2.2	1.4	1.7	0.2

This waste is at present stored as an acidic liquid in stainless steel tanks at Windscale. Although tank storage has been very successful to date, it is expensive because the tanks have to be double skinned, have duplicated cooling systems and a stirring system to prevent sludges from settling too solidly on the base of the tank. Thus for both financial and safety reasons it has been deemed preferable to solidify the waste.

3.1 The Choice of Glass. Glass has been chosen as the immobilising medium for the following reasons:

(a) It can dissolve almost all the oxides of the elements in the waste, forming a largely homogeneous body. The main parts of the waste which do not dissolve are the metals rhodium and palladium and these are safely encapsulated in the glass.

(b) Glass can be made by a relatively simple melting process which involves a minimum of handling of powders, which could increase the spread of activity in the process cell, and thus increase the radiation doses of maintenance workers.

(c) The glass can be formed at moderate temperatures (as low as 950°C, depending on the composition selected). This is useful in order to avoid wear on the components of the furnace and volatilisation of some of the waste oxides.

(d) The glass can dissolve up to 25% of the waste: a high solubility is useful to minimise the volume of solid that has to be handled.

(e) The glass can tolerate variations in the composition of the waste such as will occur from tank to tank and from top to bottom of the same tank, as sludges inevitably become concentrated at the bottom of the tank.

(f) Glass is sufficiently resistant to attack by water - this is important since dissolution of the waste form is the only credible first step in the possible return of the radioisotopes to man's environment - this will be further discussed later.

(g) Glass is stable to radiation - this is also discussed later.

(h) Glass can readily be cast into large blocks, thus again simplifying handling.

Various ceramic materials have been proposed as an alternative to glass; the most publicised is a composite of synthetic minerals known as Synroc[16]. This has not yet received as much development as glass but present indications are that, although its resistance to attack by water is better than that of glass, it will be more difficult to make, involving the handling of fine powders and the use of high temperatures (ca. 1300°C). Whether the advantages are worth a more complex fabrication procedure remains to be seen.

3.2 <u>Glass Compositions for Magnox and Fast Reactor Wastes</u>. Generally, the role of oxides in the glass structure can be divided into three classes: the 'network formers', commonly SiO_2, B_2O_3, P_2O_5, which form the three dimensional linked structure of the glass, the 'modifiers' such as the alkali and alkaline earth oxides which cause bonds in the network to be broken, and an 'intermediate' class which form some links but not in the same three dimensional manner as the network formers. These include ZnO, ZrO_2 and Al_2O_3.

Many of the waste oxides are either in the network former or intermediate class -the only important exceptions being Mg, Ba, Sr, Rb and Cs. Thus the addition of the waste to the glass does not vastly alter the network structure.

The leaching behaviour of glasses depends on the pH of the leaching medium. In acid solutions there is essentially a pure ion exchange process in which protons replace the alkali ions and the latter then diffuse to the surface. Some resistance to leaching is conferred by additions of oxides which inhibit this diffusion: typical suitable oxides are the alkaline earths and Al_2O_3, B_2O_3, and ZnO. In alkaline solutions, however, the hydroxyl ion reacts with the glass network, breaking the oxygen bridges between adjacent SiO_2 tetrahedra, producing small soluble solvate ions. In neutral solution the attack is a combination of these two processes and the rate is usually a minimum.

The original waste vitrification process, developed at Harwell in the early 1960s, was known as FINGAL. At that time, the wastes contained a much larger percentage of uranium (45%) than those from the current Magnox reprocessing plant. It was found during this programme that either borosilicate or phosphate glasses would dissolve adequate quantities of the waste, could be formed at achievable temperatures and were reasonably leach resistant[17]. However, the phosphate glasses devitrified (i.e. crystallised) easily and their leach rates then deteriorated. This can be avoided in different phosphate glass compositions, but the phosphate glasses tend to be very corrosive to plant components and receive little attention world wide at present.

For current Magnox wastes, only borosilicate glasses with formation temperatures between 900 and 1050°C have been considered. It was early found that mixed Li_2O/Na_2O glasses showed slightly improved leachabilities compared to single alkali glasses (the mixed alkali effect), with the best results somewhere near the equimolar composition[18]. Since this discovery, most of the glasses investigated here have been of the mixed alkali type.

As would be expected, increasing the alkali content or the boric oxide content at the expense of the silica increases the leach rate, but decreases the formation temperature of the glass[19]. The selection of our two reference glasses 189 and 209 is therefore a compromise between low leachability and low formation temperature. The compositions are given in Table 2.

Table 2

Glass Compositions
(in weight%)

Oxides	189 (M5)	209 (M22)
Waste	25.3	25.7
SiO_2	41.5	50.9
Li_2O	3.7	4.0
Na_2O	7.7	8.3
B_2O_3	21.9	11.1

The waste from reprocessing fuel from the Prototype Fast Reactor (PFR) is expected to contain a large proportion of Fe_2O_3 rather than the Al_2O_3 and MgO present in Magnox waste. Incorporating this waste into the glass compositions designed for Magnox waste resulted in poor leachabilities. The addition of Al_2O_3, or to a lesser extent ZnO, markedly improved the leachability, presumably, as noted above, by inhibiting alkali ion diffusion.

3.3 Constitution and Crystallisation of the Glasses. In the Magnox waste glasses, the whole of the calcined waste dissolves in the molten glass, with the exception of the rhodium and palladium. If the glass is cooled rapidly, it remains totally vitrified, although containing isolated small black flecks of these metals; the glass, however, is almost opaque, being coloured dark greeny-brown by the iron and chromium ions. However, if the glass is slowly cooled, as it will be in practice, or is heat treated in the temperature range 600-800°C, then crystals form in the glass. It is noticeable that the metallic particles act as nuclei for the other crystals.

The formation of these crystals takes place while the glass is still soft and its waste-containing properties are not affected; indeed the presence of the crystals seems to make it tougher. The leach rates of both 189 and 209 are not affected by their crystallisation. Figure 7 shows the leach rates plotted against the times that the samples had been held at temperatures of 500, 600, 700 and 800°C. This is not true for all glass compositions, however; if silica (cristobalite), or any other SiO_2-rich phase, is precipitated from a glass, then the matrix will be depleted in silica and the leach rate of the glass as a whole will increase.

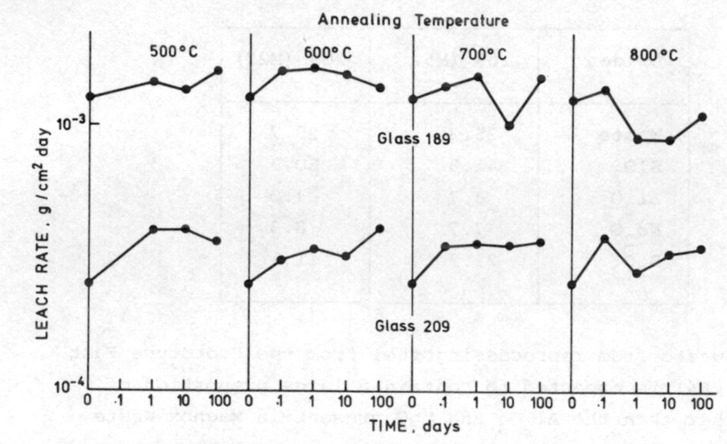

Leaching of 189 and 209 is little changed by crystallisation

Figure 7

3.4 Leaching Studies. Most of the tests that have been carried out to date on the rate of attack on the glass by water have used a large ratio of water volume to sample surface. In a waste repository on land, this condition is only likely to be fulfilled under rather improbable fault conditions. In a normal situation, after a repository has been sealed, water access will be very limited; Chapman and Savage[20] have estimated that each one tonne block of glass will 'see' only 4 ml of water per day, and even that will be saturated with silica. The rate of attack on the glass will then probably be determined by the differential solubility of silica between the temperature of the near field rock and that of the glass. (This temperature differential will be quite small by the time the can has corroded away.) Experiments to investigate this effect are in hand.

The leach rates given in Figure 7 were required largely for comparison purposes and were obtained by the Soxhlet technique, where the specimen is exposed to frequently changed, freshly distilled water at 100°C. Under these standard conditions leach rates are comparable to those obtained by the same technique for common rocks (Table 3)[21].

Table 3

Soxhlet Leach Tests at 100°C for Various Glasses and Rocks

	$mg/cm^2/day$	Age of rock
Cotswold limestone	3.4	100 MY
Jurassic limestone	2.2	
M5 glass (189)	1.8	
Bottle glass	0.3	
Arthur's Seat basalt	0.28	300 MY
M22 glass (209)	0.26	
BNWL 72-68 glass	0.22	
Pitchblende	0.1	
Window Glass	0.1	
Rockall (Eocene granite)	0.072	60 MY
Cornish Aplite	0.04	
Loch Doon granite	0.03	
Malvern rock (Gneiss-Diorite)	0.03	> 600 MY
Monazite (Ceylon)	0.03	

Experiments in flowing water have shown that:

(a) apart from an initial transition period while a layer is built up on the surface, the glasses can be regarded as dissolving (or,perhaps, being dispersed) congruently;

(b) the leach rate is very sensitive to temperature, changing by about two orders of magnitude between room temperature and 100°C[19];

(c) the impurities present in various natural waters make little difference to the rate of attack[19,21];

(d) waters of low pH have a very corrosive effect; the leach rate at a given temperature in water of pH 3 is about ten times that in pH 7;

(e) pressure per se makes little difference to the rate of

attack[22], but of course under the high hydrostatic pressures that might occur in a repository the boiling point of water is increased. If flowing water at temperatures of 200°C or more were allowed to contact the glass, the rate of attack would be quite rapid. This situation will be avoided by storing the vitrified waste with artificial cooling until the heat output from the blocks has decreased so that the maximum temperature that will be reached after disposal has fallen to a safe value or by adjusting the composition and/or spacing of the glass blocks. In addition, it is expected that the can - probably a duplex one - will protect the glass for some centuries, until the temperature has fallen to near ambient.

3.5 Radiation Stability. The glass will of course be subjected to large amounts of radiation of different kinds. The main ones are from the β/γ decays of the fission products and the α-decays of the incorporated actinides. In comparison with these two, the effects of others can be neglected - these include fission fragments and neutrons from spontaneous and induced fission in the actinides and lithium nuclei and α-particles from the $^{10}B(n,\alpha)^{7}Li$ reaction.

The γ-rays associated with the β-decays are not likely to affect the glass itself but might ionise any water in contact with it, which could possibly affect the leaching process. Most experiments designed to test this, however, have shown only small effects and leach tests on fully active samples have given similar results to tests on simulated glasses.

Most of the energy of the β-particles (electrons) will be dissipated in ionising the atoms in their path, and this is also true for α-particles, although these displace some atoms. In contrast, most of the energy of the recoiling actinide nucleus in an α-decay is dissipated in atom displacements. Ionisation will probably have only a transitory effect on the glass and the displaced atoms are the largest potential cause of changes in the glass properties.

The accumulated decays that will occur in one gram of the vitrified waste are shown plotted against time in Figure 8. Since the number of displacements produced by each α-decay is more than 4000 times the number from a β-decay (Table 4) the stability of the glass to the former has been regarded as the more important.

Table 4

Estimates of Numbers of Atoms Displaced

Assumed displacement energy		25eV	5eV*
β decay	β	0.13	1.2
	recoil	0.08	0.3
	Total	0.2	1.5
α decay	α	250	1250
	recoil	1180	5900
	Total	1400	7100

*Experiments in the electron microscope suggest that this is more nearly correct

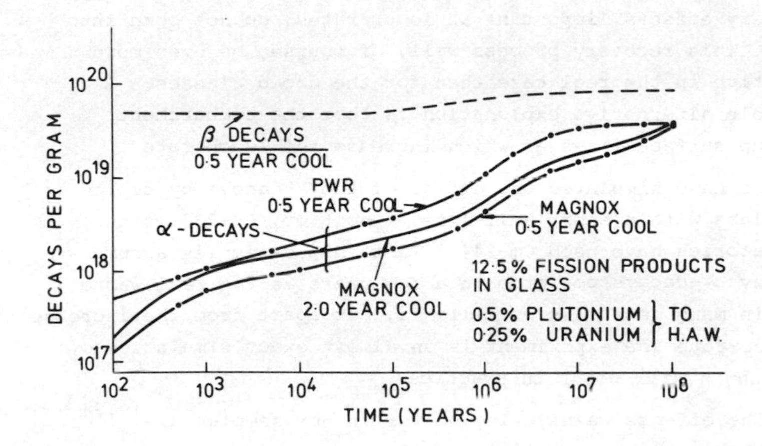

Decays per gram in vitrified waste plotted against time

Figure 8

However, the β-dose is not negligible and some experiments have been done to see if it has any effect. Most of the β-decays are from Cs-137 and Sr-90/Y-90 (90% will occur in the first 100 years). The average β energy is about 0.5MeV, giving a range in the glass of ca.1mm, and this was simulated by bombardment with 0.5MeV electrons in an accelerator, the intensity being such that the 100 year dose was reached in about one week. This resulted in slight changes in density - an 0.2% increase at 250°C and a similar decrease at 125°C - and slight increases in leach rate (ca.50%).

Because most of the displacements are caused by the recoiling nucleus it is possible to simulate the effect of the α-decays by a bombardment technique. A recently published experiment[23] used 200keV lead ions, which have a range in the glass of only 500 Å, and large increases in leach rate were observed after a critical threshold dose equivalent to ca.10^{18} α-decays per gram. However, in experiments at Harwell using the more realistic doping technique, described below, no similar increases in leachability were found after doses more than five times this threshold dose. It is concluded that the dose-rates in the bombardment technique are so high that recovery effects, important at lower rates, do not come into play. This recovery process will, of course, be even more effective in the real case than for the doped glasses. A possible alternative explanation is that the bombardment sets up surface stresses which increase the leach rate.

We have simulated the effect of the α-decays by doping the glass with a short half-life α-emitter, Pu-238; other laboratories have used Cm-244. These doped samples accumulate as many α-decays per gram in a few years as the real waste will in many centuries or millenia, but apart from the increase in dose rate the experiment is an almost exact simulation of that which will occur in practice.

The effects we have looked for in our samples are[19,24]:

(a) cracking or crystallisation:no changes were observed on microscopic examination of a thin section of the glass;
(b) stored energy: a small amount of the energy of the decays is stored by the glass (the 'Wigner' energy) but this amounts to only about 75 J g^{-1} after 1.1 x 10^{18} decays per gram. Since the specific heat is about 1 J g^{-1} K^{-1} and

since the energy is released over the temperature range of 100-450°C, no self-sustaining temperature excursion is possible;

(c) leach rates: small increases in the leach rates are observed, as is shown in Table 5, for our oldest samples of glass 189 doped with 5 wt% Pu-238.

Table 5

Leach Rates for Glass 189 Doped with Plutonium

Holding temperature:		First year / Subsequent years	50°C / 20°C	170°C / 20°C
Date	Dose (disintegrations per gram)	Equivalent time (years)	Leach rates (mg $cm^{-2}day^{-1}$) at 100°C by the Soxhlet method	
Nov 1975	0.89×10^{18}	7000	1.6	1.5
Feb 1977	2.0×10^{18}	25000	2.3	2.3
Mar 1977	2.7×10^{18}	30000	2.4	2.2
Nov 1977	2.7×10^{18}	50000	2.3	2.6
Jul 1978	3.3×10^{18}	70000	2.3	2.5
Jan 1981	5.5×10^{18}	1400000	3.2	2.7

(The equivalent times assume 0.5% Pu and 0.25%U go into the glass with the Magnox waste from fuel that was reprocessed after 6 months out of reactor; these percentages are both overestimates.)

(d) Helium release: each α-particle becomes a helium atom and it was suggested that these might form bubbles and damage the glass. No sign of this was found. Samples of glass in the form of 1cm right cylinders, or as 1mm thick discs were sealed in capsules, and held at 50°C or 170°C. No helium was released from the samples held at 50°C but, for those held at 170°C, 13% of the helium was released from the cylinders, and 50 to 85% from the discs, depending on composition. Since most of the α-decays will occur after the glass has cooled, pressurisation of the canister due to this effect is unlikely.

(e) Density changes: small density changes were observed in

a series of glasses doped with Pu-238 and held at room temperature (Figure 9); these were glass compositions suggested by various member countries of the EEC for vitrifying High Level Waste[24]. However, samples held at 170°C were found to have changed very little in density except for a French glass composition, SON 58.30.20.U2, and a phosphate glass, which showed almost the same density change as the samples held at room temperature

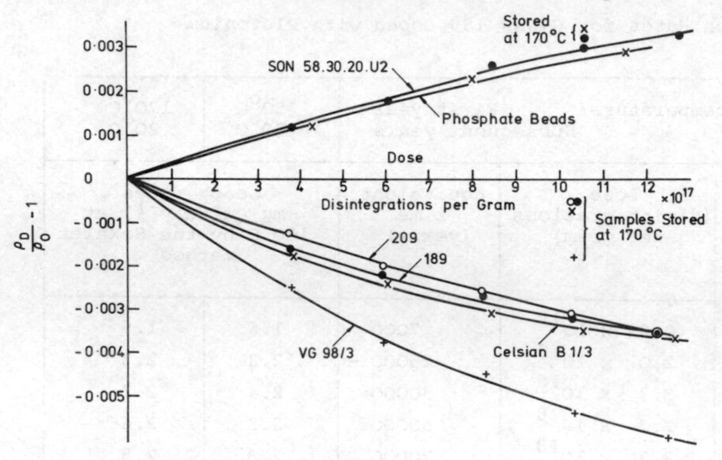

Density Changes in Glasses Doped with Plutonium-238
Figure 9

It is well known that glasses may either expand or contract under irradiation, although it is not easy to predict what will happen for a particular composition. In the present work, the composition of SON 58.30.20.U2, which contracts, is not very different from those of the other three borosilicate glasses, which expand. The density changes are showing signs of saturation, following equations of the form (Figure 9)

$$(\rho_D - \rho_o)/\rho_o = A\left[1 - \exp(-\alpha D)\right]$$

where D is the dose, in disintegrations per gram

A is the saturation value of $(\rho_D - \rho_o)/\rho_o$

$(\rho_D - \rho_o)/\rho_o$ is the fractional change in density

and α is the volume fraction of the glass totally damaged by the unit dose.

This equation can be shown to result from the assumption that the volume of glass damaged per disintegration is not further affected by any subsequent nearby disintegration which damages a volume overlapping the first. A similar form of equation results if a first order recovery process is assumed to be taking place simultaneously.

The values of the fitted constants are shown in the table:

Table 6

Fitted Constants for the Curves in Figure 9

Glass	$(\rho_D-\rho_o)/\rho_o$ at saturation	$\alpha \times 10^{17}$
189	- 0.0048 ± 0.002	0.102 ± 0.008
209	- 0.009 ± 0.003	0.04 ± 0.01
SON.58.30.20.U2	0.0065 ± 0.002	0.06 ± 0.02
VG 98/3	- 0.0089 ± 0.0008	0.09 ± 0.02
Celsian B1/3	- 0.0046 ± 0.0004	0.127 ± 0.01
Phosphate	0.007 ± 0.004	0.05 ± 0.03

It will be possible to determine these constants more accurately as the dose builds up. At present the results predict that the density changes will saturate at between 0.5 and 0.9%.

Thus we can say that these experiments all show that the glass is only slightly affected by radiation. Safety assessments of repository scenarios show that the leach resistance is adequate. Fully active half-tonne blocks of glass have been made in France for almost three years with no insuperable problems arising, and various other production processes are being studied.

In short, waste can be vitrified on an industrial scale and the resulting glass is quite satisfactory as a first generation material; it may be replaced in time by something better, but it does not seem likely to give our descendants any unexpected problems.

Acknowledgement

The authors thank Professor M. B. Waldron's group at the University of Surrey for their work on scanning electron microscopy of gel precipitated materials.

References:

[1] B.T.M. Willis. J. Physique (Paris), 25, 1964, 431

[2] M.H. Rand, R.J. Ackermann, F. Grønvold, F.L. Oetting, A. Pattoret. Rev.Int. des Hautes Temps. et Refracts, 15, 1978, 355

[3] B.T.M. Willis. Proc. Brit. Ceram. Soc., 1, 1965, 9

[4] C.R.A. Catlow. J. Chem. Soc. Faraday II, 74, 1978, 1901

[5] T.L. Markin, R.S. Street. J. Inorg. Nucl. Chem., 29, 1967, 2265

[6] T.L. Markin, E.J. McIver. Plutonium 1965 (A.E. Kay, M.B. Waldron Eds.), London, Chapman and Hall 1967, p.845

[7] R.E. Woodley. J. Nucl. Mats., 96, 1981, 5

[8] M.Tetenbaum. Rev. Int. des Hautes Temps. et Refracts, 15 1978, 253

[9] G.R. Chilton, J. Edwards. Thermodynamics of Nuclear Materials 1979. IAEA Vienna 1980, Vol. I, p.357

[10] M. Bober, O. Schumacher. Advances in Nuclear Science and Technology, 7, 1973, 121

[11] D.I.R. Norris. J. Nucl. Mats., 68, 1977, 13

[12] P.E. Potter, M.H. Rand. The Industrial Use of Thermodynamic Data. Special publication 34 (Ed. T.I. Barry) The Chemical Society, London, 1980, p.149

[13] A.S. Davidson, T.L.J. Moulding. Manufacture of the First Fuel Element Charge for PFR. Paper presented at the Symposium of the International Atomic Energy Agency, Brussels, 1973

[14] R.L. Nelson, N. Parkinson, W.C.L. Kent. UK Development toward Remote Fabrication of Breeder Reactor Fuel. Nuclear Technology, in press, 1981

[15] C.J. Hardy, E.S. Lane. Gel Process Development in the UK. Paper presented at the Symposium on Sol-Gel Processes and Reactor Fuel Cycles, Gatlinburg, 1970

[16] A.E. Ringwood, S.E. Kesson, N.G. Ware, W. Hibberson, A. Major. Nature 278, 1979, 219

[17]J.R. Grover. Glasses for the fixation of High Level Radioactive Wastes. IAEA Symposium on the Management of Radioactive Wastes from Reprocessing, Paris, 1972

[18]J.T. Dalton, A.R. Hall, L.E. Russell. AERE-R 7505, 1973

[19]K.A. Boult, J.T. Dalton, A.R. Hall, A. Hough, J.A.C. Marples in Ceramics in Nuclear Waste Management. American Ceramic Society and US DoE, Cincinnati, 1979, p.248 CONF-790420, and AERE-R 9188, 1978

[20]N.A. Chapman, I.G. McKinley, D. Savage. Proc. Workshop on Radionuclide Release Scenarios from Geological Respositories, OECD/NEA, Paris, 1980, p.91

[21]J.A.C. Marples, W. Lutze, C. Sombret, in Radioactive Waste Management and Disposal. R.Simon, S. Orlowski, Eds. CEC, Luxembourg, 1980, p.307

[22]N.A. Chapman, D. Savage, in Proceedings of Symposium on Science underlying Radioactive Waste Management. Boston, 1979, and ENPU 80-12, Institute of Geological Sciences Report, 1980

[23]J.C. Dran, M. Maurette, J.C. Petit. Science 209, 1980, 1518

[24]G. Malow, J.A.C. Marples, C. Sombret, in Radioactive Waste Management and Disposal. R. Simon, S. Orlowski, Eds. CEC, Luxembourg, 1980, p.341

Chemistry of the Nuclear Fuel Cycle

By W. L. Wilkinson
BRITISH NUCLEAR FUELS LTD., RISLEY, WARRINGTON, CHESHIRE, U.K.

1. INTRODUCTION

The nuclear fuel cycle consists of three distinct parts, viz:

(i) manufacture of nuclear fuel elements
(ii) irradiation of the fuel in power reactors
(iii) reprocessing of the irradiated fuel.

This cycle is illustrated in Figure 1.

The purpose of this paper is first briefly to review the basic chemistry underlying the processes used on an industrial scale for the manufacture of fuel elements for thermal nuclear power reactors and then to consider in more detail the chemistry involved in the reprocessing of this fuel after irradiation, paying particular attention to developments which have been made to allow the newer enriched oxide fuels to be processed.

2. NUCLEAR FUEL MANUFACTURE

The starting point for nuclear fuel manufacture in the UK is imported uranium ore concentrate. The first objective in the process, which is common to the manufacture of natural uranium metal for so-called Magnox reactors or enriched UO_2 ceramic fuel for the Advanced Gas Cooled (AGR) or light water reactors (LWR), is the production of pure UF_4.

This process consists of the following stages:-

(i) dissolution of the impure uranium ore concentrate in nitric acid, followed by filtration and purification by solvent extraction, using a solvent consisting of 20% tributyl phosphate in a kerosene diluent in a counter-current process using mixer-settlers, to give a solution of pure uranyl nitrate.

(ii) Evaporation of the uranyl nitrate solution followed by thermal denitration to give UO_3 powder in a continuous fluidised bed reactor using air as the fluidising gas.

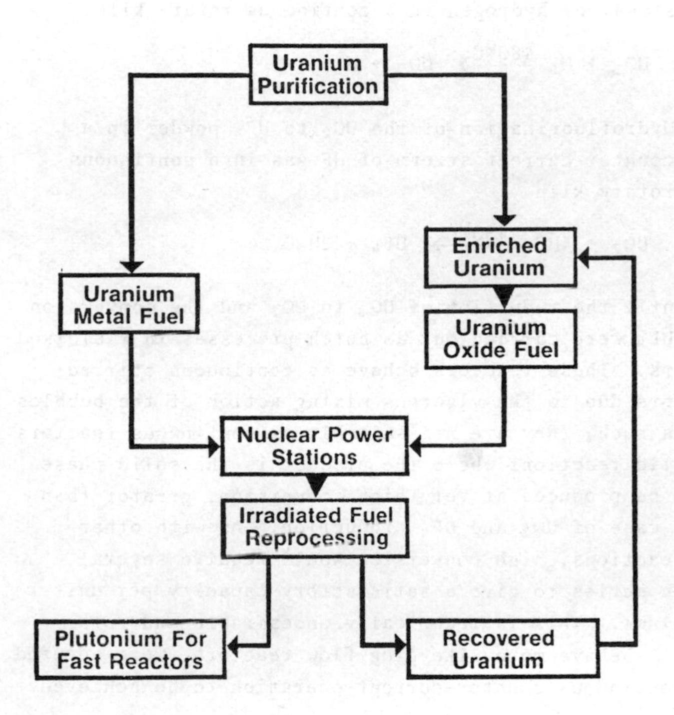

The Nuclear Fuel Cycle

FIGURE 1

$$UO_2(NO_3)_2 \xrightarrow{300^oC} UO_3 + NO + NO_2 + O_2$$

The oxides of nitrogen are removed as nitric acid by absorption in water and this is recycled to the dissolution stage.

(ii) Reduction of UO_3 to UO_2 power in a counter-current stream of hydrogen in a continuous rotary kiln

$$UO_3 + H_2 \xrightarrow{480^oC} UO_2 + H_2O$$

(iii) Hydrofluorination of the UO_2 to UF_4 powder in a counter-current stream of HF gas in a continuous rotary kiln

$$UO_2 + 4HF \xrightarrow{450^oC} UF_4 + 2H_2O$$

Until recently the reduction of UO_3 to UO_2 and the conversion of UO_2 to UF_4 were carried out as batch processes in fluidised bed reactors. These reactors behave as continuous stirred-tank reactors due to the vigorous mixing action of the bubbles of gas. As such, they are not suitable as continuous reactors for gas-solid reactions where the product is the solid phase and has to be produced at very high conversion, greater than 99% in the case of UO_2 and UF_4 production. As with other chemical reactions, high conversion would require several reactors in series to give a satisfactory capacity per unit reactor volume. This is technically undesirable and rotary kilns, which behave more like plug flow reactors, were adopted to allow continuous counter-current operation to be achieved.

2.1 Uranium metal fuel

Magnox reactors are fuelled with natural uranium metal bars, typically 3 cm in diameter and 100 cm long, clad in a finned magnesium alloy can. The uranium metal is produced from natural UF_4 by reduction with magnesium,

$$UF_4 + 2Mg \longrightarrow U + 2MgF_2$$

In this process the UF_4 powder is mixed with magnesium metal in the form of turnings or raspings and the mixture formed into pellets, typically 12 cm in diameter and 10 cm high. These pellets are loaded into a sealed graphite-lined stainless steel reactor. This is heated in an electric furnace and the highly exothermic reaction initiates at approximately 950°C to give molten uranium metal and molten magnesium fluoride slag. The uranium metal collects in a graphite mould at the base of the reactor and the slag floats on top.

After solidification the uranium billet is removed and converted into fuel element bars by conventional casting and machining processes and then canned in a magnesium alloy can with an externally finned surface to promote heat transfer.

2.2 UO_2 ceramic fuel

The fuel for the Advanced Gas Cooled Reactor, and for the two types of water reactor, the pressurised water reactor, PWR, and the boiling water reactor, BWR, consists of sintered UO_2 pellets at an enrichment of 2 to 5% ^{235}U.

The first stage in the production of such fuel is the conversion of natural UF_4 powder to gaseous UF_6 so that it can be enriched by gaseous diffusion or the centrifuge process. This conversion is carried out in a continuous fluidised bed reactor in which UF_4 powder is fluidised in a stream of fluorine, produced continuously in electrolytic cells.

$$UF_4 + F_2 \xrightarrow{450^{\circ}C} UF_6$$

The natural UF_6 is condensed and despatched to the enrichment plant.

The enriched UF_6 is returned and converted into ceramic grade UO_2 powder by means of a counter-current flow of hydrogen and steam in a continuously operating rotary kiln

$$UF_6 + H_2 + 2H_2O \xrightarrow{600^{\circ}C} UO_2 + 6HF$$

This process, which has recently been developed by BNFL, is known as the Integrated Dry Route and it is a significant advance on earlier processes which involved several stages.

The UO_2 powder produced by this process is of such a quality that it can be granulated, pelleted and sintered to give high density UO_2 pellets. These are then loaded into stainless steel cans in the case of AGR fuel elements or zirconium alloy cans for LWR fuel elements.

3. NUCLEAR FUEL REPROCESSING

The reprocessing of the spent irradiated fuel from nuclear power reactors consists of separating the fuel into its three basic ingredients. These are the unburnt uranium, the highly radioactive fission products and plutonium which is formed in the reactor from the ^{238}U isotope.

The objectives of reprocessing are twofold, viz:-

(i) to allow the radioactive fission products to be isolated and converted into a form, eg by conversion into glass, in which they can be safely stored and eventually disposed of away from man's environment in deep geologic formations, on the deep ocean bed or in deep ocean sediments.

(ii) to give access to a further very important source of energy, plutonium, which in a total recycle system involving thermal and fast reactors can give over 60 times the energy which a given quantity of uranium would give in a once-through thermal reactor system.

The basic method used for reprocessing spent nuclear fuel was originally developed to deal with metallic uranium fuel. It comprises dissolution of the fuel in nitric acid, purification from fission products and separation of uranium and plutonium by the Purex solvent extraction process employing tri-butyl phosphate as extractant in an inert diluent, followed by thermal denitration of the uranyl nitrate product to uranium trioxide and precipitation and calcination of the plutonium to its dioxide. This process is illustrated in Figure 2. This same basic method will be used in the THORP plant at Windscale to reprocess oxide fuels of much higher fission product, plutonium and other actinide contents, arising from LWR and AGR nuclear power programmes.

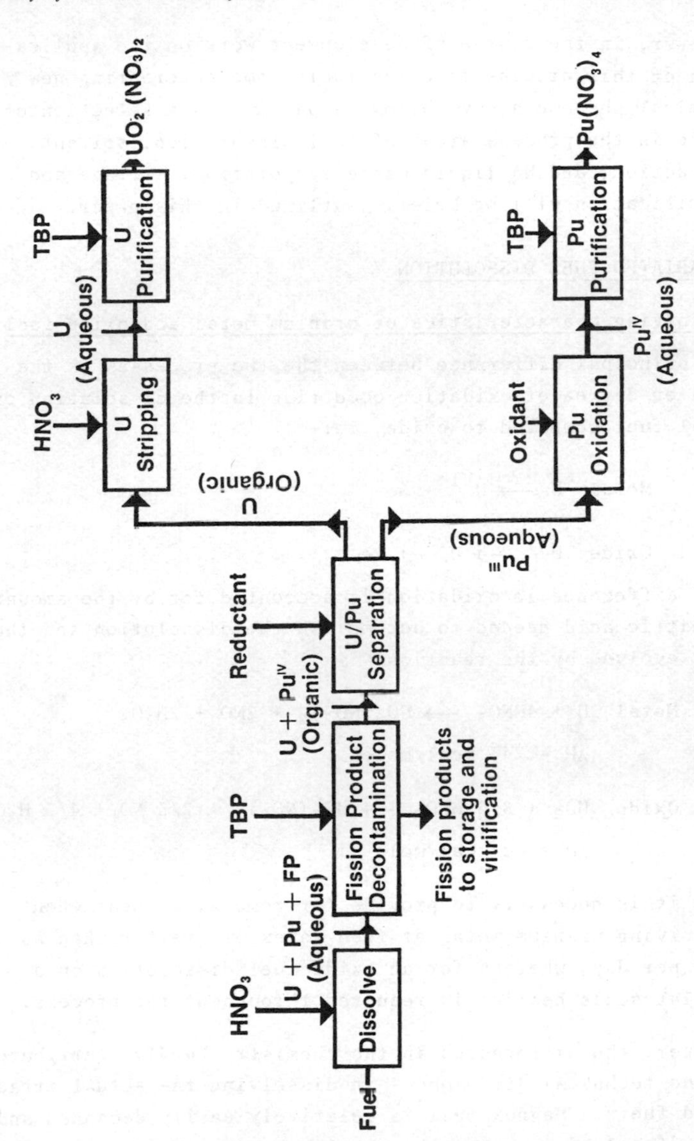

Irradiated Fuel Reprocessing

FIGURE 2

However, in the course of development work on the application of this process to oxide fuels some interesting new chemical phenomena have become apparent and a selection of these in the process areas of fuel dissolution, solvent extraction, and HA liquid waste evaporation, storage and vitrification will be briefly outlined in this paper.

4. IRRADIATED FUEL DISSOLUTION

4.1 Dissolving characteristics of uranium metal and oxide fuel

The principal difference between the two processes is the greater degree of oxidation occurring in the dissolution of metal fuel compared to oxide, ie:-

Metal $U \longrightarrow U^{VI} + 6e^-$

Oxide $U^{IV} \longrightarrow U^{VI} + 2e^-$

This difference in oxidation is accounted for by the amount of nitric acid needed to accomplish the dissolution and the heat evolved by the reaction.

Metal $U + 4HNO_3 \longrightarrow UO_2(NO_3)_2 + 2NO + 2H_2O;$

$\Delta H = 245$ kcal/mole

Oxide $UO_2 + 8/3\ HNO_3 \longrightarrow UO_2(NO_3)_2 + 2/3\ NO + 4/3\ H_2O;$

$\Delta H = -24$ cal/mole

Thus it is necessary to provide for removal of heat when dissolving uranium metal at feed rates of greater than about 5 te per day, whereas for an oxide fuel dissolution on a similar scale heating is required throughout the process.

However, the differences in the chemistry hardly contribute to the technical differences in dissolving the actual irradiated fuels. Magnox fuel is relatively easily decanned and then dissolved in a continuous dissolver at Windscale. This process requires no solids removal and only low off-gas flows are experienced, whereas LWR fuel is much more difficult to declad and the alternative of bulk shearing of fuel without decanning favours the use of a batch dissolver.

This results in high off-gas flows and problems associated with the removal and monitoring of the can debris (hulls) from the dissolver.

4.2 Dissolver solids from uranium metal and oxide fuels

Only a very small amount of insoluble fission product particles are formed in the dissolution of metal fuel. The particles are fine, typically 1-2 microns or less, and are predominantly palladium, probably as the oxide PdO, although other fission products Ru, Rh, Mo, Zr are also present along with some plutonium. At the modest temperatures and irradiation levels of metallic fuel insufficient migration of fission products occurs to enable the formation of many particles large enough to be insoluble in nitric acid solution.

In contrast, for oxide fuels, a significant amount of insoluble material arises (0.1-0.3% by weight for thermal fuel) containing the fission products Ru, Rh, Mo, Te, Zr, Pd. Ruthenium and molybdenum are the major components by weight. A wide range of particles sizes has been seen, depending on the fuel irradiation and reactor temperature conditions, from less than 0.1 micron up to 5-10 microns and occasionally a few hundred microns. Particle shape is also variable, spheres being the most common, but rod-like or needle-like particles and occasionally even more complex shapes have been seen. The occurrence of these solids can be explained in broad terms from the higher irradiation achieved in oxide fuel generating more fission products and the higher temperatures enabling elements to migrate in the fuel to form insoluble particles. Since upon fission a general oxygen deficiency results, the more noble fission products are able to form metallic alloys. One other factor that cannot be excluded is that upon dissolution some fission products approach or exceed their solubility in nitric acid and precipitation of additional solids could occur.

The insoluble material from oxide fuel dissolution has to be removed and the intention is to use solid-bowl centrifuges for this purpose. An alternative possibility is to take advantage of the fact that fission product insolubles are paramagnetic and make use of the technique of High Gradient Magnetic Separation. In this the slurry flows through a column containing an open structure, a ferromagnetic matrix, and surrounded by a high flux magnet. When the magnetic field is on the paramagnetic particles are attracted to the matrix and allow the clarified liquid to proceed. When the magnetic field is switched off the particles can be flushed out. This device offers attractions in fuel reprocessing since, unlike a centrifuge, it is mechanically simple and suitable for maintenance-free conditions.

4.3 Iodine chemistry during the dissolution of metal and oxide fuels

The differences in iodine behaviour between the dissolution of metal and oxide fuel arise not as a result of any fundamental difference in dissolution chemistry of the fuels but from the different dissolver techniques employed. In each case fission product iodine exists in the fuel as iodides of other fission products, particularly caesium, but upon dissolution it is rapidly oxidised to molecular iodine.

Because the Magnox cladding can be removed without undue difficulty the continuous dissolution process noted earlier is used for the uranium metal rods which are fed into a refluxing nitric acid solution from which a continuous overflow is taken to the solvent extraction process. In the boiling nitric acid environment oxidation of the iodine is prevented by the action of nitrogen oxides evolved by the dissolution reaction and by nitrous acid in the nitric acid solution. The molecular iodine is volatile and would escape in the off-gas from the Windscale dissolver but for the fact that the off-gas is scrubbed counter-currently by the feed of fresh nitric acid entering the dissolver. In this way a high proportion of the iodine originally in the fuel is held in the dissolver solution as molecular iodine and passes into the solvent extraction section.

A continuous dissolver for LWR fuel is less readily employed and in THORP it is proposed that sheared LWR fuel will be fed in batches to hot nitric acid held in kettle-type dissolvers.

Once again the iodine is readily oxidised to molecular iodine which, being volatile, escapes with the other off-gas. In this case there is insufficient reflux liquor to return much of the iodine to the dissolver and the bulk is lost from the dissolver solution in this way. With the completion of feeding of sheared fuel the dissolution reaction subsides and the dissolver solution is raised to boiling in order to complete the dissolution of the remaining traces of fuel. During this reflux phase iodine remaining in the dissolver will tend to be oxidised to iodate and thus volatilisation of iodine will decrease.

5. SOLVENT EXTRACTION CHEMISTRY

The reprocessing of nuclear fuels by solvent extraction, based on the Purex process, is a well proven and tested technology. However, the reprocessing of oxide fuels from advanced thermal and fast reactor systems presents some new and interesting problems. These arise mainly as a result of three factors, namely tighter product specifications and the industry's self-imposed stringent environmental discharge limits, coupled with the higher fuel burn-up achieved in advanced reactor systems.

The higher burn-up and tighter product specifications mean that much higher decontamination factors have to be achieved across the solvent extraction cycles. The nature of the problem can be seen in the Table below where the decontamination factors required for the major species in the projected oxide fuel reprocessing plant at Windscale (THORP) are compared with those specified for the present Magnox fuel reprocessing.

Decontamination Factors required for Magnox & Oxide Fuel Reprocessing

Decontamination Factor	Magnox	Oxide
Ru	1×10^6	1×10^8
Np	1.0	5×10^4
Pu	1×10^5	5×10^8
Tc	1.0	1×10^3

In the Purex process the irradiated nuclear fuel, which contains both fission products and transuranic elements in addition to the original uranium, is dissolved in nitric acid. The U and Pu are then extracted together into the tributyl phosphate in kerosene diluent (TBP/OK) in order to separate them from the bulk of the fission products as indicated in Figure 2. The Pu and U are separated from each other by reducing the plutonium from the extractable tetravalent state to inextractable trivalent state, thus directing the plutonium to the aqueous phase away from the uranium-bearing solvent phase. The plutonium and uranium are then further purified from each other and from fission products in the purification cycles. These processes can be carried out in pulsed columns or mixer-settlers or combinations of the two.

In this paper three examples will be given of recent developments to solve the problems associated with high burn-up fuels in the solvent extraction process. These are:-

(a) the use of U(IV) in the form of uranous nitrate as a reducing agent for plutonium,

(b) the behaviour of neptunium,

(c) the behaviour of technetium.

5.1 Chemistry & kinetics of plutonium reduction

In order further to reduce effluent discharge levels it is important to develop processes which are 'salt-free' in order to allow maximum concentration factors for medium active waste to be achieved without crystallisation occurring and thereby minimising storage costs. This has involved new reagents such as hydrazine-stabilised uranous nitrate and hydroxylamine nitrate as reducing agents in place of the ferrous sulphamate which has been used in the past.

The principle of the separation of uranium and plutonium in a pulsed column contactor (Figure 3) is as follows. Uranium and plutonium from the HA cycle enter column 1 in the solvent phase at a point near the bottom of the column. Also present in this solvent feed are nitric acid and some nitrous acid.

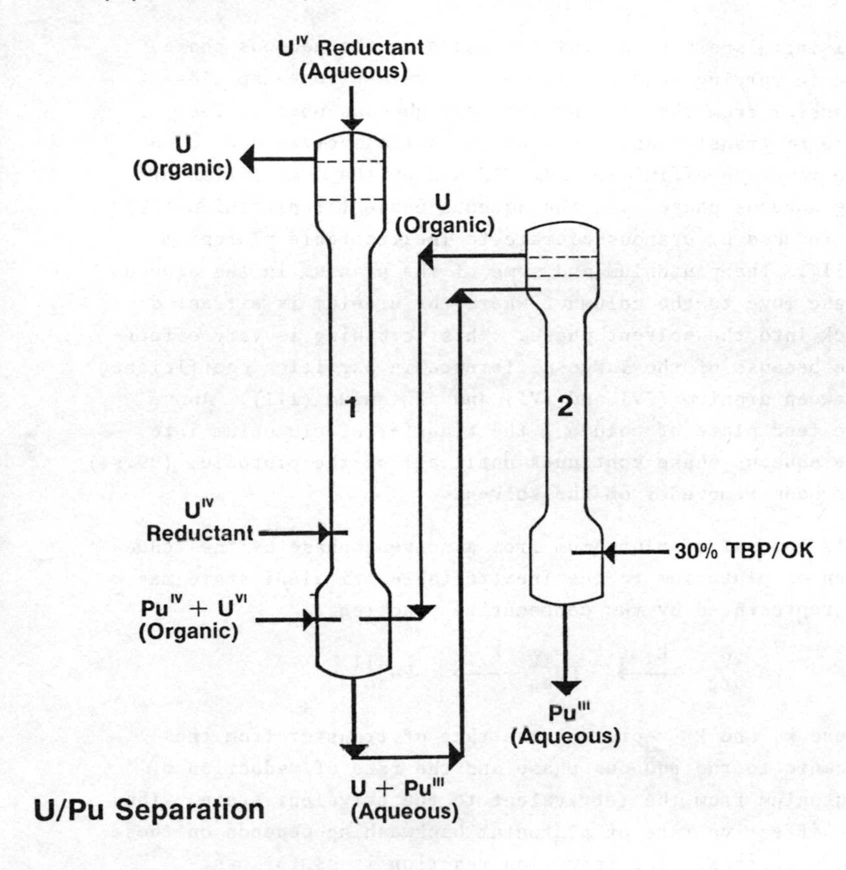
U^{IV} Reductant
(Aqueous)
U
(Organic)
U
(Organic)
1
2
U^{IV}
Reductant
30% TBP/OK
$Pu^{IV} + U^{VI}$
(Organic)
Pu^{III}
(Aqueous)
$U + Pu^{III}$
(Aqueous)
U/Pu Separation

FIGURE 3

Pulsing disperses the solvent phase in the aqueous phase and in varying degrees all four above-mentioned species transfer from the solvent into the aqueous phase. The rate of transfer and the quantities transferred depend on the relative affinities for TBP and on the composition of the aqueous phase. In the aqueous phase the plutonium (IV) is reduced by uranous nitrate to inextractable plutonium (III). The plutonium and some of the uranium in the aqueous phase move to the column 2 where the uranium is extracted back into the solvent phase. This scrubbing is very effective because of the large difference in partition coefficient between uranium (IV) and (VI) and plutonium (III). Above the feed plate of column 1 the transfer of plutonium into the aqueous phase continues until all of the plutonium (99.9%) has been removed from the solvent.

This removal of plutonium from a solvent phase by the reduction of plutonium to the inextractable trivalent state may be represented by two consecutive reactions:-

$$Pu^{IV}_{org} \xrightarrow{k_1} Pu^{IV}_{aq} \xrightarrow{k_2} Pu^{III}_{aq}$$

where k_1 and k_2 represent the rate of transfer from the organic to the aqueous phase and the rate of reduction of plutonium from the tetravalent to the trivalent state. Thus the effective rate of plutonium backwashing depends on these two processes. The reduction reaction is as follows:-

$$2Pu^{4+} + U^{4+} + 2H_2O \rightleftharpoons 2Pu^{3+} + UO_2^{++} + 4H^+$$

Perhaps the single most important variable in the reduction of plutonium (IV) by uranium (IV) is the effect of aqueous phase nitric acid concentration. It has been shown that the rate of reduction is inversely proportional to the square of the hydrogen ion concentration. It is advantageous therefore to operate the plutonium backwash section under as low an acid concentration as is possible since the reaction will proceed much faster. The acidity of the aqueous phase column 1 is determined largely by the acidity of the solvent product from the preceding scrub stage and by the solvent to aqueous ratio in column 1.

The effect of the U(IV)/P(IV) molar ratio on the performance of column 1 has also been studied. In addition to ensuring that sufficient uranium (IV) is fed to the column to bring about the reduction of all the plutonium (IV) to plutonium (III), it is also necessary to ensure that the uranium (IV) is distributed in a suitable profile along the column length.

The type of contactor chosen in which to carry out the separation may also influence the rate of plutonium backwashing. The control of the operation of a contactor and of the physical parameters to ensure a favourable kinetic regime is easier in a discrete stage contactor, such as a mixer-settler, than in a differential contactor such as a pulsed column. However, because of the high plutonium throughput, pulsed columns have been chosen for use in the HA cycle of THORP and will be used for the splitting stage.

The effects of reductant concentration, contact time and nitric acid profile on the performance at this stage of the process have been examined in a series of experimental runs carried out on pilot scale pulsed columns.

5.2 The behaviour of neptunium

The transuranic element neptunium-237 is produced from uranium fuel in nuclear reactors by the following reactions:

$$^{238}U\ (n,\ 2n) \rightarrow {}^{237}U \xrightarrow[6.7d]{\beta} {}^{237}Np$$

$$^{235}U\ (n,\gamma) \rightarrow {}^{236}U\ (n,\gamma)\ {}^{237}U \xrightarrow[6.7d]{\beta} {}^{237}Np$$

Approximately 70% of the ^{237}Np is produced by the first reaction.

Because of the long half life of ^{237}Np (1.7 x 10^6 years) the input of neptunium to the reprocessing plant rises steeply with irradiation, and with fully irradiated PWR fuel it may be expected to approach 500 g Np/te U. This has to be compared with the 17 g Np/te U arising from the present generation of Magnox reactors. It is important to maintain neptunium in the uranium product at a low level since the uranium has to be recycled via the enrichment process. If the specification for Np in the uranium product is to be met for oxide fuel then a decontamination factor of 1 x 10^4 will be required: the decontamination factor achieved in the Magnox plant at present is 100-250.

General Chemistry of Np

The basic chemistry of neptunium has been established for many years. In nitric acid the (IV), (V) and (VI) valency states are commonly encountered. Valency changes are slow unless accelerated by an increase in temperature or the presence of HNO_2, and these factors render the behaviour of Np during reprocessing difficult to predict. The problem is that the solvent extraction of neptunium in TBP systems is very complex. The tetra and hexavalent states are extractable whilst the pentavalent stage is virtually inextractable (partition coefficient $< 10^{-2}$). Therefore in both strong oxidising and strong reducing conditions neptunium is extracted by TBP and if good decontamination factors are to be achieved conditions favouring the formation of the pentavalent form are essential.

Tetravalent Np

In nitric acid Np (IV) is only stable in the presence of reducing or complexing agents, and in the absence of these an equilibrium of (IV), (V) and (VI) will exist. The potential of the Np (IV)/(V) couple is 0.74V; thus reducing agents with a potential of around 0.54V, eg N_2H_4, should produce solutions having 99.9% Np (IV). This reaction is very dependent on nitric acid strength and by selection of suitable conditions redox reagents with higher potentials can be successfully used (eg Fe^{2+}/Fe^{3+}, 0.77 V).

The reduction of Np (VI) to Np (IV) is a two-step process. Np (VI) to (V) is fast whereas Np (V) to Np (IV) is much slower.

Pentavalent Np

In nitric acid Np (V) is stable only at low acidities (0.2-1M), the rate of disproportionation increasing with acidity until 10M when Np (V) completely disappears.

Np (V) oxidation by NO_3^- is slow and incomplete at normal acidity (1-3M). However, NO_2^- has a significant effect on the position of equilibrium and the rate of reaction. In trace concentrations NO_2^- functions as a catalyst for the oxidation of Np (V) to Np (VI) by NO_3^-. At higher NO_2^- concentrations (> 10^{-2} M) the nitrite itself acts as a reducing reagent and prevents oxidation of Np (V). Np (V) may be reduced to Np (IV) at ambient temperatures, at low acidities, by the use of N_2H_4(slow) or Fe^{2+} (fast).

Hexavalent Np

Np (VI) is highly stable in concentrated nitric acid and the lower valency states can be oxidised to the hexavalent state under boiling conditions. Reduction to Np (V) can be achieved by use of Fe^{2+}, $NH_2OH.HNO_3$, N_2H_4, U (IV) or HNO_2; however only $NH_2OH.HNO_3$ will give a rapid and exclusive yield of Np (V), the other reagents tending to produce Np (IV).

The pattern of changing valency states during the dissolution and conditioning cycle can be followed from absorption spectra. These demonstrate the importance of changes in nitrous acid concentration. On completion of dissolution, when the nitrous acid concentration is very low, the Np is present as Np (VI). In order to reduce Pu (VI) present after dissolution, conditioning of the dissolver liquor is carried out with nitrous acid leading to a concentration of about 0.05-0.1 M HNO_2. Under these conditions Np is present largely as Np (V). After the conditioning the excess nitrous acid is removed by sparging with air leaving a final nitrous acid concentration of about 1 x 10^{-3} M. Under these conditions most of the neptunium, but not all, is present in the inextractable pentavalent state.

There are several ways of eliminating this neptunium from the system. One technique is to arrange conditions so that it follows the uranium stream in the U/Pu splitting stage and is then reduced to the pentavalent state so that it can be effectively removed from the uranium stream in its inextractable form in the uranium purification section.

5.3 The behaviour of technetium

The fission product Tc^{99} is produced in thermal fission of U^{235} by a decay chain with one of the highest fission yields (about 6.3% of the total fission product yield from U^{235} fission and 6.1% of the yield from Pu^{239} fission). Tc^{99} is a low energy beta emitter and is one of the longest lived fission products with a half life of 2.1×10^5 years. It is significant therefore from an environmental point of view.

The concentration of Tc^{99} in irradiated fuel is directly proportional to burn-up and for Magnox and PWR fuel the quantities are approximately as follows:-

Fuel	Burn-up	Tc g per te U
Magnox	5000 MWD/te U	137
PWR	36,000 MWD/te U	940

Dissolution of fission product technetium (metal or oxide) by nitric acid forms the pertechnetate anion. This species is weakly extracted into tributyl phosphate solutions according to the equilibrium

$$H^+ + TcO_4^- + (3TBP)_{org} \rightleftharpoons (HTcO_4\ 3TBP)_{org}$$

In the presence of certain metal ions, however, co-extraction of technetium occurs. Uranium is extracted from nitrate solutions by tributyl phosphate as $UO_2(NO_3).2TBP$ but in solutions containing technetium a mixed complex $UO_2(NO_3)(TcO_4).2TBP$ is formed. The complex is formed by replacement of one nitrate ligand by pertechnetate according to equilibrium.

$$[UO_2.(NO_3)_2.2TBP]_{org} + TcO_4^- \rightleftharpoons [UO_2.(NO_3)\ (TcO_4).2TBP]_{org}$$

Thus, like ruthenium, zirconium and niobium, technetium is partially extracted into the organic phase during the initial extraction cycle of nuclear fuel reprocessing. Reprocessing plants presently in use have been unaffected by technetium, partly because its specific activity is so low, that high DFs are not required in order to reduce its radiation levels to acceptable values. This is not the case with high burn-up oxide fuels.

Partition data indicate that at acidities of 3-4N HNO_3, and in the absence of uranium in the extraction section, the extraction of Tc will be low. However, in the scrub section at acidities of 2-3N the extraction will be high and DFs low. The partition coefficient for Tc is about five times that of Np under scrub conditions; hence it follows that it may be difficult to achieve high DFs in conditions that promote a high DF_{Np}, ie high U saturation and low acidity which are precisely those that lower DF_{Tc}. Hence initial assessments indicated that the U product was at risk.

However it became evident that, in the presence of reducing agents, partition coefficients of Tc are low and in fact the Pu product specification was at risk. The flowsheet has therefore to be carefully tuned.

Technetium gives rise to a further complication in U/Pu separation. In the U/Pu splitting stage the reflux of uranium within the columns 1 and 2 of Figure 3 continues until equilibrium is reached and uranium leaves 1 in the solvent. Ideally no plutonium would be refluxed in the columns. However, the trivalent plutonium is not stable and under certain conditions varying amounts of it are oxidised and reflux does take place. As oxidation reactions increase refluxing increases leading to increased plutonium concentrations throughout the U/Pu separation column. Unacceptable separation occurs when the plutonium reflux increases until the reducing agent can no longer maintain the plutonium in the trivalent state. Some plutonium(IV)would then remain in the organic phase as it leaves the top of column 1.

Oxidation of trivalent plutonium by nitrate ion is slow in the absence of a catalyst. However, nitrous acid which is present in equilibrium amounts in many of the process streams and is produced in situ by radiolysis of nitric acid is an effective catalyst for the oxidation reaction. Nitrous acid concentration is therefore controlled by the addition of hydrazine. It is clear therefore that any reaction that destroys hydrazine is detrimental to the stability of the U/Pu separation system.

It has been found that hydrazine decomposition occurs in the presence of Tc and that the destruction of hydrazine is far more than can be accounted for by redox reaction with technetium. The significance of this is that if this destruction of hydrazine takes place in the U/Pu splitter column then the control of nitrous acid concentration at that step of the process will be difficult.

It is therefore important to understand the factors which affect the $Tc/N_2H_4/HNO_3$ reactions and the effects of U (IV), P (IV), nitric acid concentrations and temperature have been studied so that satisfactory operating conditions can be defined and maintained.

6. TREATMENT OF HIGHLY ACTIVE RAFFINATES

6.1 Evaporation and storage of HA liquors

As HA raffinate is evaporated at 5 cm Hg pressure and $60^{o}C$, solids are precipitated. Analysis of the solids in the HA liquor from metal fuel reprocessing shows that these consist mainly of zirconium, molybdenum, phosphorus and caesium. Evidence from laser-Raman measurements suggest that the principal species are zirconium phosphate, and caesium phosphomolybdate, with possibly some zirconium molybdate, zirconia and molybdenum oxide.

The main difference in the solids arising in oxide fuel HA concentrate is the ratio of fission products to phosphate. The total solids precipitated are about the same but in different proportions. The oxide concentrates give a smaller volume of solids for a given concentration of fission products, and this has led to the idea of storage of the HA concentrate in a simplified form of tank, with little or no agitation. In such a tank the concentration would be limited by the heat output of the settled solids and the depth of sludge. However, due to a recent decision to commence vitrification of the liquid wastes at an early date it is more economical to store HA concentrates from oxide fuels in the existing type of tank rather than develop new designs.

In the concentrate liquors from metallic fuels, the main crystallising species are the nitrates of non-active elements Mg, Al and Fe. (Magnesium comes from adherence of residual traces of cladding metal to the declad fuel metal, aluminium and iron are incorporated in the fuel rod to improve its metallurgical properties in the reactor and iron also arises as a corrosion product). However, results from recent active work show that lanthanides are also present in the crystallised solids, probably forming co-crystals with magnesium. This is a consequence of reprocessing higher burn-up fuel.

In oxide concentrate liquors there is a much higher proportion of fission products (the level is proportional to the burn-up) and lower levels of fuel additives. Experimental work has shown a number of different crystallising species (Sr/Ba, La/Nd/Ce and Gd) and the precise crystallising species are dependent on irradiation levels, corrosion products and the presence of soluble neutron poisons, notably gadolinium.

6.2 HA Waste Vitrification

The ultimate disposal of the highly active liquid wastes arising from nuclear fuel reprocessing requires their fixation in a stable matrix and BNFL have chosen borosilicate glass for this purpose.

The process for the solidification of HA liquid wastes comprises the following stages:-

(i) Conditioning or Pretreatment. During storage the concentrated HA liquid wastes from reprocessing at Windscale are kept acidic in order to minimise precipitation of phosphatic and hydroxide solids. It is also possible to condition this liquor prior to vitrification in order to reduce the volatility of Ru in the subsequent processes. Conditioning can take the form of acid destruction or neutralisation.

(ii) Evaporation. Free acid and water are boiled off leaving a residual slurry of concentrate and suspended solids. The temperature range is 100-200°C.

(iii) Calcination. The waste components are partially decomposed to the constituent oxides in the temperature range 350-800°C.

(iv) Melting. In this stage the products of calcination are mixed with glass frit and the waste constituents are incorporated into the glass matrix at temperatures up to 1300°C.

(v) Disposal. Requires a casting step to produce a solid block in a canister or container which is then stored.

In the process to be used for the vitrification of Windscale wastes, steps (i), (ii) and (iii) will be carried out simultaneously in a rotary kiln calciner.
The calcined waste will then pass directly into a melter where, at a temperature of about 1100°C it will be incorporated into the glass matrix.
The rotary kiln consists of a slightly inclined rotating tube, in either a resistance or induction heated furnace. The highly active liquid waste is pumped in at the raised, colder end and in its passage along the length of the tube, where the temperature progressively increases, the wastes are dried and decomposed to the oxides. The calcined product is then blended with glass frit and the mixture fed

to the top of the melter. In the melter, at a temperature of about 1100°C the waste oxides are incorporated into the molten glass. Periodically the solid plug of glass in the bottom outlet of the melter is melted to allow the molten glass to flow into the container. Off-gases from the calcining and melting stages are routed to a clean-up system employing filters, scrubbers and absorbers.

<u>Thermal decomposition of wastes</u>

The decomposition of batches of highly active liquid waste has been studied in a thermobalance and the thermal decomposition curve of a simulated liquid waste shows that the waste begins to lose weight at 36°C continuing very slowly up to 80°C. From this point to about 200°C, where an inflection appears, the weight loss is rapid. This is mainly due to evaporation of the free water and nitric acid in the waste and of products from the decomposition of compounds which break down at comparatively low temperatures, eg nitrates of Al, Cr and Fe.

From about 200°C where the next inflexion appears the rate of loss increases slowly to about 420°C at which temperature the almost complete decomposition of nitrate compounds except those of Ba, Cs and Sr might be expected. The weight loss at this temperature is nearly 90%. There is then a slower weight loss up to a temperature of about 526°C. It is considered that at this temperature all the nitrate compounds except those of Ba, Cs and Sr and rare earths have decomposed completely.

Above 560°C the wastes continue to lose weight very slowly up to 915°C at which temperature a constant weight is reached. This is due to complete decomposition of the remaining nitrate compounds.

However, in the dynamic situation prevailing in the continuous rotary kiln the extent of the decomposition reactions described above depend both on the temperature and the residence time in each of the calciner heating zones. Also the chemistry is complicated by the addition of various reagents, notably $LiNO_3$, to the process.

$LiNO_3$ is added to the calcine to render elements with a tendency to form refractory oxides, eg Al, Fe and Cr, more reactive. The reaction between $LiNO_3$ and the refractory metal oxides is probably a solid phase reaction since the decomposition temperature of $LiNO_3$ is higher than the peak temperature usually encountered in the calciner.

ACKNOWLEDGEMENT

The author is grateful for the considerable help he has received from G D C Short in the preparation of this paper.

Chemistry in the Service of Electricity Generation

By D. J. Littler
RESEARCH DIVISION, CENTRAL ELECTRICITY GENERATING BOARD, COURTENAY HOUSE, 18 WARWICK LANE, LONDON EC4P 4EB, U.K.

1. The Historical Background

The association of chemists with electricity goes back, perhaps, as far as Joseph Priestley[(1)], the discoverer of oxygen. In 1775 he commented on the importance of the inter-relationship between electricity and chemistry, also rather dryly noting that "few of our modern electricians have been either speculative or practical chemists". He was inaugurating the switch of scientific interest from electrostatic to chemical generation of electricity, but it was the physicist Volta, who, with his chemical pile, sounded what Davy, who was certainly a chemist, recognized as "an alarm bell to experimenters all over Europe".[(2)]

When electricity first became available in significant amounts to the experimenter it was, of course, the chemist who was responsible for its generation because it was produced entirely from batteries and electrochemical cells. Before very long the primary cells, as we should now call them, were producing very large electric currents, which enabled chemists to take first steps in the preparation of metals, electrochemistry and in the demonstration of electroplating.

Still in the era of electricity from electrochemical cells, the first regular journal on electricity appeared in 1833. It acknowledged the role of chemistry in the title "Electricity, Magnetism and Chemistry" and was concerned with improvements in primary cells. Its cover is shown in Plate 1.

By the 1830's Faraday's demonstration that a voltage could be produced in a wire driven across a magnetic field had opened the way for the conversion of fuel energy into electricity via rotating machines without the complexities of chemical processes and electro-chemical cells and the modern era of electricity generation was born. At this point the chemist began to give way to the physicist and engineer in the electricity

THE ANNALS

OF

ELECTRICITY,

MAGNETISM, & CHEMISTRY;

AND

Guardian of Experimental Science.

CONDUCTED BY

WILLIAM STURGEON,

Lecturer on Experimental Philosophy, at the Honourable East India Company's Military Seminary, Addiscombe, &c. &c.

AND ASSISTED BY GENTLEMEN EMINENT IN THESE DEPARTMENTS OF PHILOSOPHY.

VOL. I.—OCTOBER, 1836, TO OCTOBER, 1837.

London:

Published by Sherwood, Gilbert, and Piper, Paternoster Row; and W. Annan, 12, Gracechurch Street.
Sold also by Messrs. Hodges and Smith, and Fannin and Co. Dublin; Maclachlan and Stewart, and Carfrae and Son, Edinburgh; Mr. Robertson, Glasgow: Mr. Smith, Aberdeen; and Mr. Dobson, No. 108, Chestnut Street, Philadelphia.

1837

PLATE 1

generating scene.

Public supplies of electricity appeared in several advanced countries in the 1880s, the first in the United Kingdom being commissioned in 1881 at Godalming. Generation was powered by a water wheel in the river Wey which is the river running below this University. An idea of what the system actually contributed to the night-life of Godalming is given by the contemporary print reproduced in Plate 2.

During the ensuing struggle of the infant electricity supply industry, the chemist took on quite a different role. He became active in applying the techniques of chemical analysis to aid the emergent companies with their fuel purchasing, in the control of the purity of water used in their boilers, in introducing the concept of the control of efficiency of conversion of the fuel energy into electrical energy and also in environmental monitoring and control - all themes which have continued to the present era.

Wilhelm Ostwald, the Father of Physical Chemistry and a great chemical thermodynamist, pointed the road to the pre-occupation of chemists with efficiency of fuel-to-electricity conversion when he said[3] "I do not know whether all of us realize fully what an imperfect thing is the most essential source of power which we are using in our highly developed engineering - the steam engine". Ostwald was, of course, looking back nostalgically to the primacy of the chemist in the days of the primary battery which in the eyes of the chemist was intrinsically capable of far greater conversion efficiency. The possibility of converting primary fuel directly into electricity in what became known as the fuel cell has since those days remained of great interest to chemists and has attracted an heroic research effort. It has not yet produced a wholly economic competitor to the use of rotating machinery, however.

From the 1920's onwards the chemist became increasingly important in electricity generation, assisting in the smooth and economical operation of equipment, in devising new operational techniques and setting higher standards for future plant. In this he was helped by new methods of chemical analysis, especially those for surface examination, interactions at surfaces playing a large part in energy conversion processes. There have also been great developments in the science of combustion catalysis,

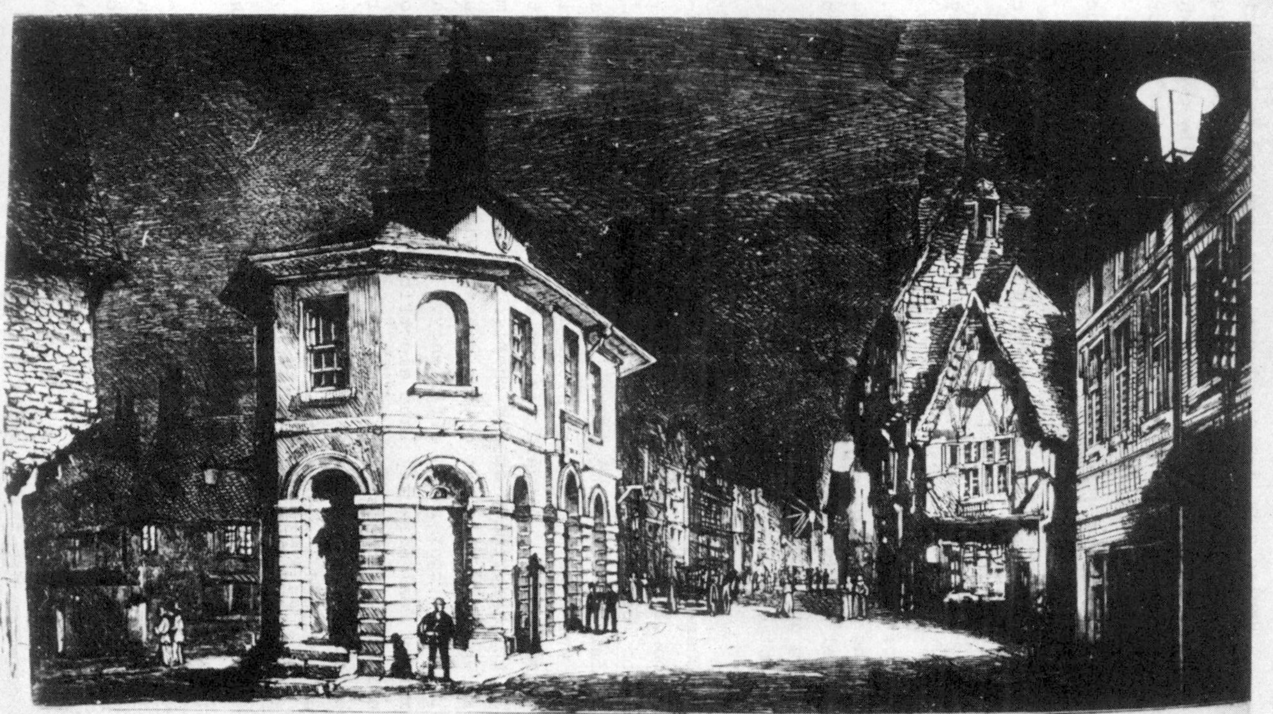
488 THE GRAPHIC Nov. 12, 1881

THE TOWN OF GODALMING ILLUMINATED BY THE ELECTRIC LIGHT

corrosion, the chemistry of the solid state and environmental monitoring.

2. Chemists and the CEGB

In 1947, an important organisational change took place within the U.K.'s electricity supply industry. The Electricity Bill received the Royal Assent on 13th August of that year and subsequently, on vesting day, 1st April, 1948, the British Electricity Authority (BEA) and fourteen Area Electricity Boards became responsible for the public system of electricity supply throughout Great Britain, with the single exception of the north of Scotland. This change was not without significance for the employment of chemists and other professional scientists and engineers, directly within the industry.

With only 12900 MW installed capacity the unified industry was quite small by today's standards, but it provided an industrial base large enough to support the growth of an in-house research effort to tackle the emerging issues of reliability, performance and environmental protection for the large new generating stations then being planned.

Research was further stimulated by the 1956 report of the Herbert Committee,[4] following its inquiry into the industry. This recommended a substantial increase in research expenditure, then running at approximately £¾M per annum. By 1979-80 this had expanded to £64M, while the industry itself had grown to some 57000 MW installed capacity. Today the Central Electricity Generating Board (CEGB), successor to the original BEA, employs over 250 professional chemists and chemical engineers in its three central research laboratories and five regional Scientific Services Departments (SSDs). Each of the 130 generating stations also has its small chemical laboratory and Station Chemist.

Station Chemists are responsible for the routine chemical aspects of day-to-day operation. Their colleagues in the SSDs deal with operational problems which require deeper investigation and also do basic research allied to such problems. The central laboratories have wider scope. With their extensive facilities and expertise, they are able to tackle major problems and can often offer support to the regional laboratories. They also investigate possible future developments.

I shall select three areas in which chemists have

made particularly important contributions. These are:

1) work on the mechanisms and control of fireside corrosion in coal and oil fired boilers.

2) the chemistry of the interactions of carbon dioxide coolant with steel fuel cladding and graphite in advanced gas cooled reactors.

3) boiler-water chemistry.

I shall not dwell upon the chemical work being carried out on the environmental impact of electricity generation. This will be covered in the following paper.

3. Fireside Corrosion in Coal and Oil Fired Power Stations

3.1 Thermal Efficiency. The efficiency with which the heat content of fuels is converted to electrical energy is central to the economy of electricity generation. In steam raising power stations the efficiency of conversion is controlled by the value of the highest temperature to which the steam is heated and also by the lowest temperature at which heat is rejected. The latter is determined by the temperature of the surroundings e.g. the rivers or the sea into which heat is rejected. The highest temperature is that of the superheaters (see Figure 1) from which the steam passes to the inlet to the turbines. Limiting factors on increasing the superheater temperature are the strength and corrosion resistance of the steels used to construct the turbines, tubes and ducting. Materials strength lies within the field of the metallurgist, but the chemist has made a major contribution to identifying the processes whereby steels are corroded by the hot fluids (steam and flue gas) with which they come into contact.

3.2 Superheater Corrosion Problems

3.2.1 Coal-Fired Power Stations. The main cause of fireside corrosion in coal-fired boilers is 'impurity' in the fuel. As supplied, power-station coal can contain up to 20 wt% of mineral matter[5]. Much of this is in the form of relatively unreactive alumino-silicates, which are converted to ash during the combustion process. At the same time, however, a number of other elements, such as sulphur, sodium and chlorine, are mobilised and subsequently form aggressive deposits on exposed tube surfaces.

The average power station coal contains 1.6wt% sulphur,

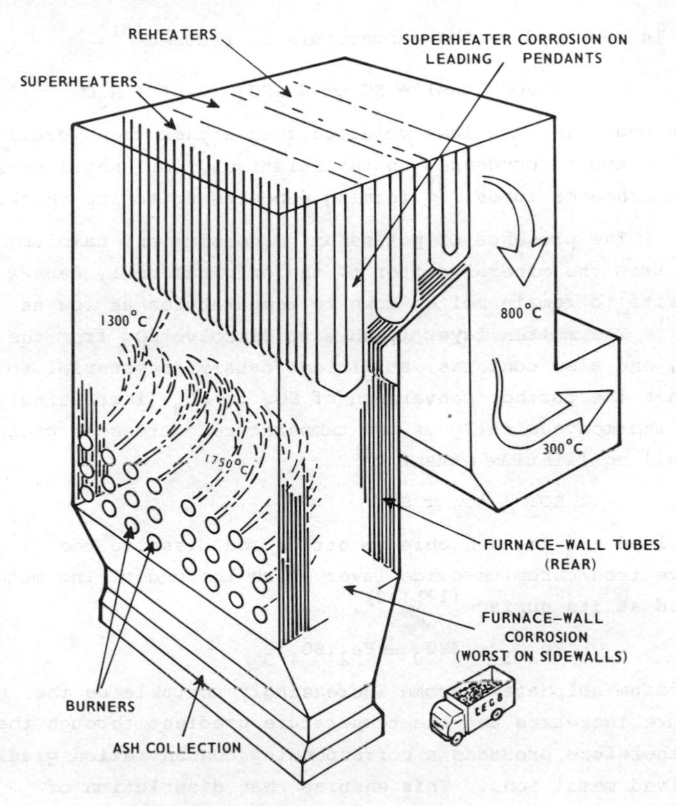

FIG. 1 SCHEMATIC DIAGRAM OF A FRONT-FIRED BOILER, THE REGIONS MOST SUSCEPTIBLE TO FIRE-SIDE CORROSION BY THE COMBUSTION GASES AND ASH PARTICLES ARE INDICATED

about half of which is directly bonded to the organic matter, the rest being distributed as fine particles of iron pyrites, FeS_2. During combustion, the sulphur is oxidised to SO_2, a small amount (approximately 1%) being further oxidised to SO_3 by the excess oxygen (2 to 4 vol%) in the gas stream[(6)]. Sodium averages 0.15wt% in coal, occurring in both the alumino-silicates and as adsorbed cations in the organic matter[(7)]. It vaporises in the flame, and then reacts with other components of the combustion gases (e.g. SO_3, H_2O and HCl), the latter being the main combustion product of the chlorine in the coal. Coal contains on average 0.26 wt% chlorine, primarily as adsorbed anions in the organic matrix.

At combustion temperatures (1500°C+), NaOH and NaCl are the preferred vapour phase species, but conversion to the sulphate is favoured as the temperature is reduced[8]:

$$2NaOH\ (gas) + SO_3 \rightleftharpoons Na_2SO_4\ (gas) + H_2O$$

Sodium sulphate is much less volatile than either the hydroxide or chloride and it condenses on the relatively cool metal surfaces of the superheater tubes,[9] forming deposits up to 1mm thick.

The presence of potassium, magnesium and calcium, released from the mineral matter of the original coal, causes the deposits to remain molten down to temperatures as low as 600°C.[10] The molten layer is able to dissolve SO_3 from the flue gas, and also contains sufficient catalytic material to ensure that the further conversion of SO_2 to SO_3 (increasingly favoured thermodynamically as the temperature decreases) proceeds to its full equilibrium extent[11].

$$SO_2 + \tfrac{1}{2}O_2 \rightleftharpoons SO_3$$

The dissolved SO_3 is then able to attack and dissolve the protective iron/chromium-oxide layer which the underlying metal has formed at its surface[12][13].

$$Fe_2O_3 + 3SO_3 \rightleftharpoons Fe_2(SO_4)_3$$

Iron/chromium sulphates become increasingly unstable as the temperature increases and the temperature gradient through the deposit therefore produces a corresponding concentration gradient of dissolved metal ions. This ensures that dissolution of protective oxide is effectively a continuous process, with iron and chromium diffusing outwards through the molten sulphate. At the hotter outer surface they are reprecipitated, but no longer form a protective layer.

It follows that corrosion will be at its most severe at those points on a tube where the heat flux (and hence the temperature gradient) is greatest, and this is found to be the case. Fig. 2 shows the characteristic pattern of a corroded superheater tube, with the main zones of corrosion being at the 2 and 10 O'Clock positions on the circumference relative to flue gas flowing from 12 O'Clock. The aerodynamics of the gas flow cause entrained ash particles to build up a protective layer at 12 O'Clock, thus restricting the heat flux at this position. The ash is much thinner around the sides of the tube (Fig. 2b), where the heat flux is consequently greater[10].

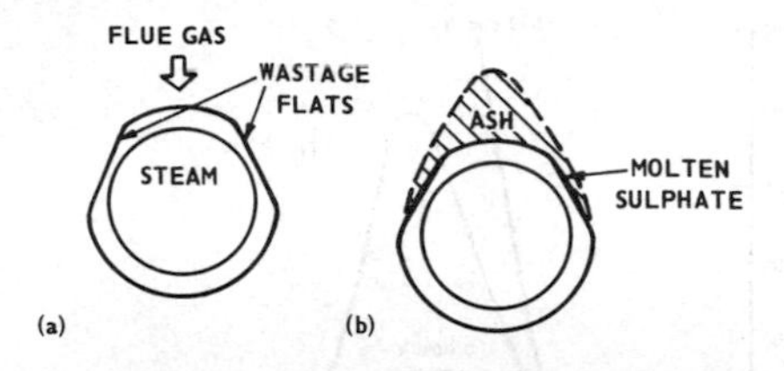

FIG. 2 SUPERHEATER CORROSION PROFILES

(a) CLEAN TUBE CROSS-SECTION
(b) IN-PLANT CROSS-SECTION SHOWING DEPOSITED MATERIAL

3.2.2 Oil-Fired Power Stations. Superheater corrosion in oil-fired boilers follows a general pattern similar to that in coal-fired units, although the distribution of fuel impurities is very different. Residual fuel oil contains only 0.1 to 0.4wt% alumino-silicate minerals, but between 2 and 4wt% sulphur; the sodium content generally lies between 50 ppm and 300 ppm, with between 50 ppm and 200 ppm vanadium.[7] During combustion, the vanadium (present in the fuel as a component of large porphyrin molecules) is oxidised to involatile V_2O_4,[14] which forms a fine fume in the gas stream.[15] As the gases cool, however, residual oxygen in the flue gases is absorbed to cause partial conversion to V_2O_5.

$$V_2O_4 + \tfrac{1}{2}O_2 \rightleftharpoons V_2O_5$$

and subsequent absorption of sodium leads to the formation of vanadates. It is from this fume, together with unabsorbed sodium converted to sulphate in the flue gases, that the superheater deposits are formed.[16] The deposits absorb SO_3, and the vanadates may themselves play a direct role in fluxing away protective oxide from the metal surface. The corrosiveness of the deposit is sensitive to its sodium-to-vanadium ratio. This is illustrated in Fig. 3, which shows the corrosion rates of an 18Cr 12Ni Nb steel exposed to combustion gases generated from oil containing 50 ppm sodium and various levels of vanadium.[11] Also shown, for comparison, are the oxidation rates for the steel in air.

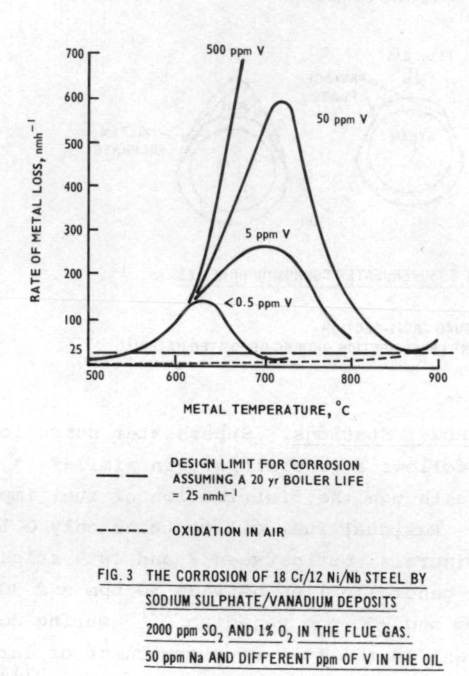

FIG. 3 THE CORROSION OF 18 Cr/12 Ni/Nb STEEL BY SODIUM SULPHATE/VANADIUM DEPOSITS
2000 ppm SO_2 AND 1% O_2 IN THE FLUE GAS.
50 ppm Na AND DIFFERENT ppm OF V IN THE OIL

3.2.3 Application of Chemical Data to Superheater Design.
The data which the chemists have provided have given the CEGB a firm basis from which to optimise operational conditions for superheaters in both coal- and oil-fired boilers. The target is to limit corrosion rates to an average of no more than 25 nm h^{-1}, so giving a full 20 y life to the 8 mm wall tubes used in these boilers. With this in mind, a maximum steam temperature of 565°C, (corresponding to a tube outer surface temperature of approximately 650°C) has been fixed by the CEGB for coal-fired plant. For oil, the corresponding temperatures are 540°C and 580°C, reflecting the more aggressive nature of the deposits produced by this fuel.[10]

This does not mean that all tubes are expected to achieve this lifetime. Leading tubes in a bank can experience conditions far more arduous than those within the bank and may well require replacement after intervals of only 4 to 6 y. Even here, however, lifetimes are well beyond the two years which separate the statutory overhauls on operational boilers, so that badly corroded tubes can be identified and replaced well before they fail. From the economic point of view, this is immensely important; if a 500 MW coal-fired boiler has to be brought off load because of a tube failure during a period (say, mid-winter) when its maximum output is required, generating replacement power with less efficient standby plant can incur additional costs of up to £100,000 per day with a minimum of two days per outage. These costs must subsequently be reflected in the cost of electricity. The national importance of the chemists' contributions, both in this and the water chemistry field (see Section 5), was recognised by the Royal Society in 1977, when they presented CEGB chemists with the ESSO Award for Conservation of Energy.[17]

One final point on superheater corrosion; it will be noted (Fig. 3) that the corrosion rates for the different deposits go through maxima and then decline at higher temperatures, eventually reaching rates no greater than those produced by oxidising the metal in air alone. The decline reflects the reduced solubilities of SO_3 and the metal oxides in the molten deposits at these temperatures, and it might be thought that this could be exploited to provide higher thermal efficiencies. However, problems primarily related to maintaining the strength of the tube material at these temperatures make this unattractive.

3.3 Furnace-Wall Corrosion in Coal-Fired Boilers. Over the last few years the main emphasis of fireside corrosion studies has changed from the superheater to the mild-steel tubes which line the walls of the boilers close to the burners; their locations are shown in Fig. 1. The metal here is much cooler (only 400°C to 450°C); in normal, oxygen-rich flue gas it forms a protective oxide scale and normally corrodes tolerably slowly. In recent years, however, a number of boilers have suffered particularly severe corrosion, sufficient, in the worst cases, to cause failure of a new tube well within two years. The Board's chemists and materials specialists are now investigating the reasons for this, and how it might best be prevented[18]. In a significant early step, they demonstrated that the combustion gases sweeping the worst-affected areas of tubing contained significant amounts of only partially oxidised material (e.g. up to 10 vol% CO, as well as carbon grits). These reducing conditions promote the formation of sulphides within the oxide scale, and since these have greater molar volumes than oxide they cause mechanical disruption, making the scale less protective and hence promoting the further oxidation of the underlying metal.[10] The reasons for the locally high CO and carbon almost certainly lie with the kinetics of coal-particle combustion in combination with the gas flow dynamics of the boiler. This finding has opened the way to investigating how modifying the burner firing patterns might reduce the impingement of the still burning gases on the furnace wall. It is worth noting that this is not a problem in oil-fired boilers, where the combustion of fuel oil is more rapid and efficient. Although reducing conditions may occur locally, they are not sufficiently severe to promote corrosion through sulphide formation.

A particularly important observation from the work is that severe corrosion can generally be associated with the use of high-chlorine coal. British power-station coals have exceptionally high chlorine contents (averaging 0.26%Cl, with 30% containing more than 0.30% and 5% more than 0.5wt%) and it is with the higher-than-average chlorine levels that the worst difficulties have been linked.

Much remains to be settled regarding the details of the chemistry by which the chlorine might be affecting the corrosion process. In broad outline, however, it appears that the combustion product HCl further disrupts the oxide layer. It may also participate in the transport of metal from the tube surface as volatile chlorides (e.g. $FeCl_2$).[19] To combat this, tube metals are being investigated which form surface oxides more resistant than mild steel. Examples are alloys rich in chromium, such as 25%Cr20%Ni and 50%Cr50%Ni.

An alternative approach is to see whether chlorine can be removed from coal prior to burning. In collaboration with the NCB, Board chemists have shown that the chlorine is distributed fairly uniformly throughout the coal substance, adsorbed on the walls of the micropores which permeate its structure.[20] The extractability of the chlorine thus depends on the accessibility of these micropores. This can vary substantially, as demonstrated by water-leaching experiments on a series of 3 to 6 mm crushed coals.[21] These showed that while 60 to 70% of the chlorine could be removed from some coals, less than 20% was released from others. Board chemists are currently investigating ways whereby chlorine accessibility in difficult coals might be improved, at acceptable costs.

4. Gas-Side Problems in Nuclear Power Stations

4.1 The Gas-Cooled Reactor. Over 10% of the power generated by the CEGB is from nuclear plant,[22] and this proportion is expected to grow steadily over the next two decades. Chemists played a central role in nuclear power development and have made important contributions to the commissioning and operation of the Board's nuclear power stations.

The Board currently operates two types of reactor, in both of which the heat is extracted from the core and transferred to the steam-raising equipment by circulating carbon dioxide gas. From a chemical point of view, the main difference between the two types lies in the temperatures to which the gas stream is heated in the core - up to 360°C (for a fuel can maximum of 400°C) in the earlier Magnox types and up to 650°C (for a top fuel temperature of 830°C) in the advanced gas-cooled reactor (AGR) series. Gas pressures are also different; in Magnox stations they range from 10 atm at Bradwell to 27 atm at Wylfa, while in the AGRs they are 30 atm for Dungeness B and 40 atm for the others. Figure 4 shows the basic layout of the AGR circuit.

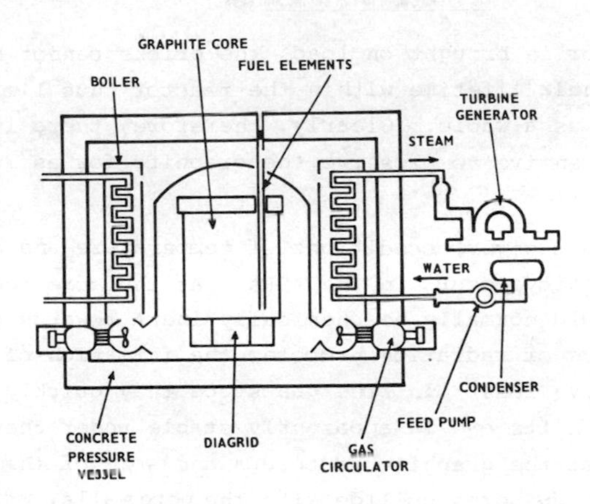

FIG. 4 THE AGR CIRCUIT

In operating either type of reactor, a critical balance has to be maintained between oxidation and reduction processes in the core - oxidation and loss of the graphite moderator on the one hand, and reduction of the CO_2 coolant to form carbon deposits on the fuel cladding on the other. The Board's chemists, in collaboration with colleagues in the UKAEA, have played a fundamental role in showing how this balance may be achieved and maintained.

4.2 Oxidation of Core Graphite. Each reactor core is constructed of a matrix of keyed graphite moderator bricks into which the fuel elements are inserted (see Fig. 5)

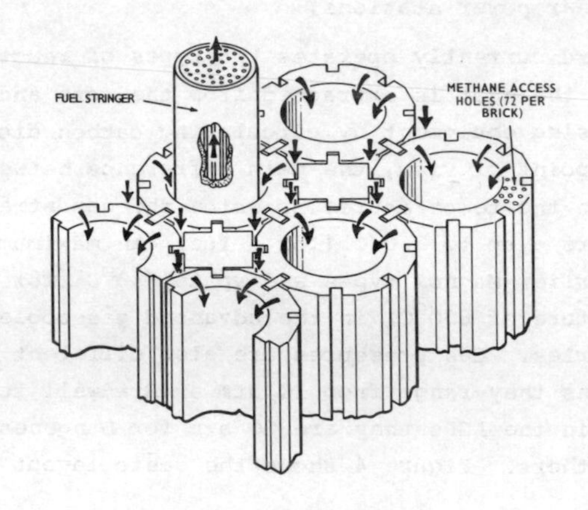

FIG. 5 CAGR MODERATOR BRICKS AND KEYS

Once the reactor is brought on load, the bricks cannot be replaced and their lifetime within the reactor thus limits that of the reactor as a whole. Clearly, therefore, there is a substantial incentive to preserve the graphite for as long as possible.

The most severe conditions of temperature and nuclear radiation intensity occur in the AGRs. At the core temperature ($400^{o}C$) CO_2 would normally be chemically inert towards graphite, but the presence of radiation promotes the formation of highly reactive positive ions. In free gas space they quickly recombine and hence CO_2 on its own is apparently stable under these conditions. But the graphite is porous and some of the ions produced within the pores collide with the pore walls, which they oxidise and convert to gas. This leads to a progressively increased porosity and consequent weakening of the bricks. Unchecked, the process would eventually cause the bricks to fracture.[23] Figure 6 shows the relationship between brick strength and degree of oxidation.

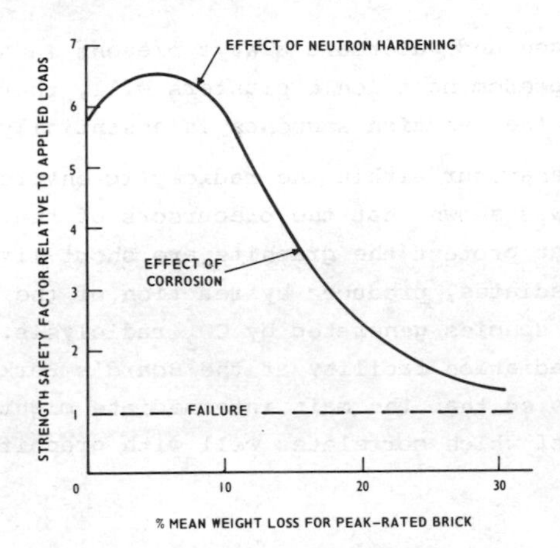

FIG. 6 VARIATION OF BRICK STRENGTH WITH GRAPHITE CORROSION

The problem was appreciate early in the development of the AGR and solutions were sought. The presence of CO, naturally formed in small quantities in the coolant gas, was known to be able to influence the rates of both graphite oxidation and carbon deposition, while subsequent research by the UKAEA demonstrated that methane was a very powerful corrosion inhibitor. A few hundred ppm are sufficient to stop the gasification reaction almost completely[(24)] as the radiolytic decomposition of the methane results in the deposition of a sacrificial layer of carbon on the graphite surfaces. Unfortunately, carbon can also deposit from the methane onto the fuel, impairing heat transfer and leading to unacceptable increases in fuel temperature. The question thus arose as to how the protective ability of these species might best be exploited.

The first step was to establish what the basic mechanisms of radiolytic oxidation of graphite by CO_2 are. An important question was the identity of the intermediate CO_2 irradiation products which actually react with the graphite surfaces. This was a matter of debate for several years before mass spectrometric studies, sponsored by the Board at Liverpool University, demonstrated that they are probably clustered positive ions of the general type $(CO_2)_n^+$. It was also shown that the presence of CO inhibited their reactivity by replacing CO_2 in the ionic clusters,[(26)] e.g.

$$(CO_2)_3^+ \xrightarrow{CO} (CO_2.CO.CO)^+ \xrightarrow{CO} (CO)_3^+$$

In practice, hydrogen and water are always present in AGR coolants, and the predominant ionic clusters will, therefore, be protonated, but the reaction sequence is essentially similar.

Methane behaviour within the radiolytic environment was also studied. It was shown that the precursors of the carbonaceous deposits that protect the graphite are short-lived hydrocarbon intermediates, produced by reaction of the methane with the oxidising species generated by CO_2 radiolysis. Evidence from the gamma-irradiation facility at the Board's Berkeley Nuclear Laboratories suggested that the main intermediate might be ethane, the concentration of which correlates well with graphite corrosion rates (Fig. 7).

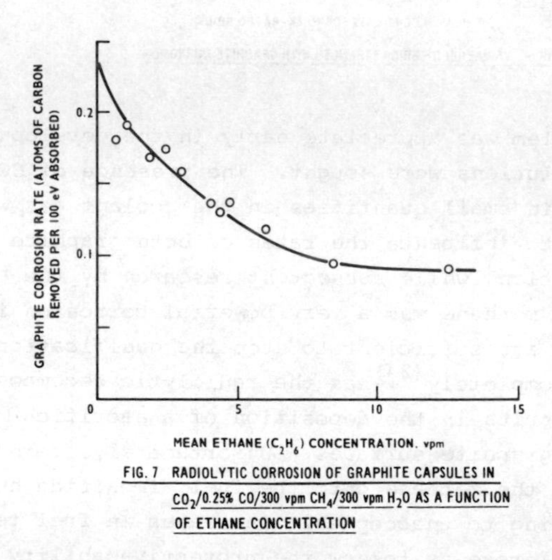

FIG. 7 RADIOLYTIC CORROSION OF GRAPHITE CAPSULES IN CO_2/0.25% CO/300 vpm CH_4/300 vpm H_2O AS A FUNCTION OF ETHANE CONCENTRATION

Ethane itself could not be injected directly into the reactor, as it would decompose before reaching the areas it was meant to protect, but if methane is injected instead, ethane is actually generated by radiolytic reactions in the free gas spaces of the core in close proximity to just those areas.

The efficiencies with which CO and methane inhibit graphite oxidation are strongly influenced by the graphite pore structure and the data acquired have allowed the corrosion process to be modelled in terms of these factors.[27] This has since played an important part in defining operational conditions for the AGRs and, in particular, has assisted in the development of an improved design of AGR core, the 'ventilated core'. In this the gas-leakage paths are modified to force the irradiated gas stream through the graphite bricks, rather than through the gaps between them, so ensuring that all the vulnerable parts are exposed to inhibitor; the net result will be an increase of several years in core life. The new design is being exploited in the UK's latest AGRs, Heysham II and Torness.

4.3 Carbon Deposition on Fuel-Cladding Surfaces. Core-graphite lifetimes might be extended if carbon deposition from the coolant could be increased; the constraint on this, as mentioned above, is the need to minimise carbon deposition on the steel fuel cladding. Currently, the problem is handled in operational AGRs by maintaining tight control on the levels of CH_4 (100 to 200 ppm) and CO (0.5 to 2 vol%) so that both oxidation and deposition stay within design limits over the whole of the reactors' 30-year design life; see Figure 8.

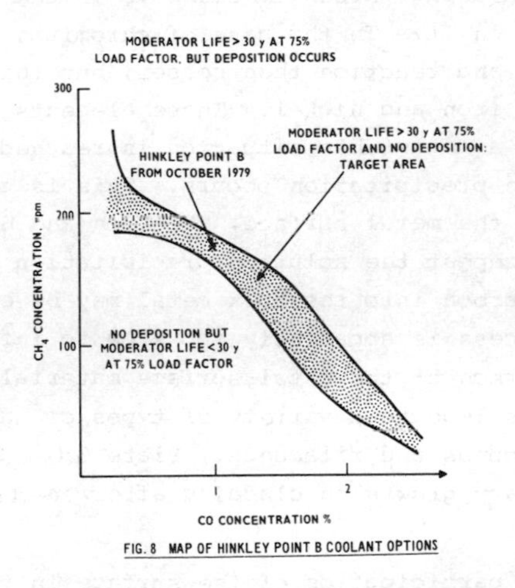

FIG. 8 MAP OF HINKLEY POINT B COOLANT OPTIONS

However, the fuel is changed on average every five years, so it is reasonable to ask whether it would be possible to take advantage of a more graphite-protective coolant, while still keeping carbon build-up on the cladding within tolerable limits during this much shorter period. There are two possible lines of approach; the first is to adjust the chemistry of the gas so as to inhibit deposition selectively at the relatively high temperature of the cladding surface; the second is to modify the nature of the surface itself so as to slow down its reaction with the gas phase.

At high temperatures, it has been found that deposition

increases with the proportion of unsaturated hydrocarbons, (such as ethylene formed from methane[28]) in the coolant gas, and that this proportion itself increases as the temperature is raised. One way of reducing deposition might, therefore, be to suppress the formation of these unsaturated hydrocarbons, possibly by increasing the level of hydrogen in the coolant. To exploit the alternative approach of modifying the cladding surface, it is important to understand the reactions taking place between the gas phase and the various components of the steel. Studies have shown that the metal surface plays an important part in the overall deposition process, with significant differences in behaviour between its three main constituents, iron, nickel and chromium.[29][30] Carbon build up is a two-stage process, in the first of which anyone of the three elements reacts with the gas-phase species, causing carbon to dissolve in the metal surface rather than deposit on it. In the case of chromium, stable carbides are formed and reaction then ceases, but this does not appear to be so for iron and nickel. These elements continue to absorb carbon until a metastable situation is reached and, as a second stage, carbon precipitation occurs. This is the beginning of carbon growth on the metal surface, the iron and nickel now being liberated to repeat the solution/precipitation cycle. Some diffusion of carbon into the bulk metal may be expected to occur, but this process is apparently too slow to influence the precipitation of carbon at the metal surface materially. Different conditions lead to a variety of types of carbon growth, including plates, mounds and filaments. Plate 3 (p. 211) shows an example of filamentary growth on cladding after reaction in CO at 550°C.

The active participation of the surface in the process suggests that deposition might be inhibited if the fuel cladding could be provided with an inactive surface layer. A number of possibilities are being researched; one is to deposit a very thin layer of inert silica by the decomposition of an organo-silicon compound. Development of this type of coating has now reached the point of in-reactor testing. The potential for using chromium rich surface layers is also being investigated. Figure 9 illustrates the resistance to carbon deposition of fuel cladding coated with a 50μm layer of chromium, when held at 800°C in CH_4. The behaviour of an uncoated control sample is also shown. The results are promising, though much still remains to be

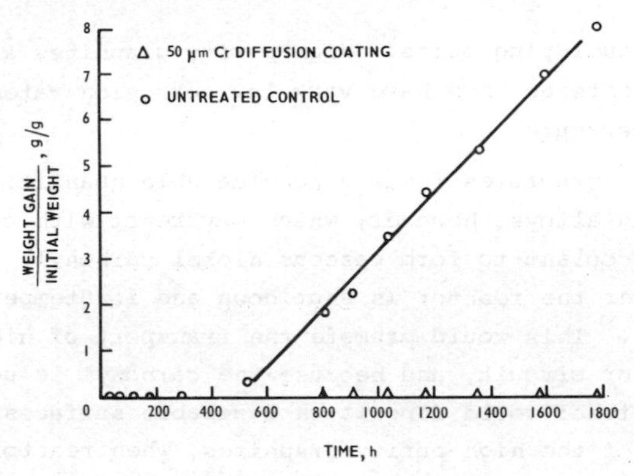

FIG. 9 CARBON DEPOSITION ON CHROMIUM COATED 20/25 Nb FUEL CAN CLADDING IN CH_4 AT 800°C, 1.67 Atm

settled before adequate oxidation resistance and adhesion to the substrate metal can be assured.

4.3 Other Gas-Side Problems. The chemists' contribution to gas-side issues is not confined to elucidating and controlling reactions which occur in the core. They have made important contributions to a wide variety of other topics including:

i) the establishment of new and improved techniques for maintaining potentially difficult adventitious impurities such as sulphur and fission product I^{131} at acceptable levels in the coolant.[31][32]

ii) the development of a novel type of photoionization detector which gives an order-of-magnitude improvement on the previous detectability of oil leaked into the CO_2 from the gas circulators.[33]

iii) the resolution, in collaboration with materials specialists, of the processes whereby boiler steels are oxidised in hot CO_2.[34]

A particularly interesting example, from the chemist's point of view, has been the work on the thermal oxidation of AGR graphites. The bulk of reactor graphite is used in the core, where, at temperatures in the region of 400°C, radiolytically induced corrosion predominates, but there are several graphite components which experience the maximum gas temperature (650°C) and whose oxidation is controlled by thermal processes. Examples are the graphite fuel sleeves and the piston seals and bearings of the fuel-plug units. High-purity graphites are used for these components as they have very low corrosion rates in CO_2 at this temperature.

Reactor structures contain considerable quantities of nickel containing alloys, however, which may react with carbon monoxide in the coolant to form gaseous nickel carbonyl ($Ni(CO)_4$) whenever the reactor is shut down and low temperatures (<250°C) pertain. This would promote the transport of nickel around the reactor circuit, and because the carbonyl is unstable, finely divided nickel would deposit on available surfaces, including those of the high-purity graphites, when reactor power is raised once more. Laboratory research[35] has shown that the corrosion of these graphites in $CO-CO_2$ coolants is strongly catalysed by nickel. This is demonstrated in Figure 10 which relates the yearly oxidation of graphite samples to their levels

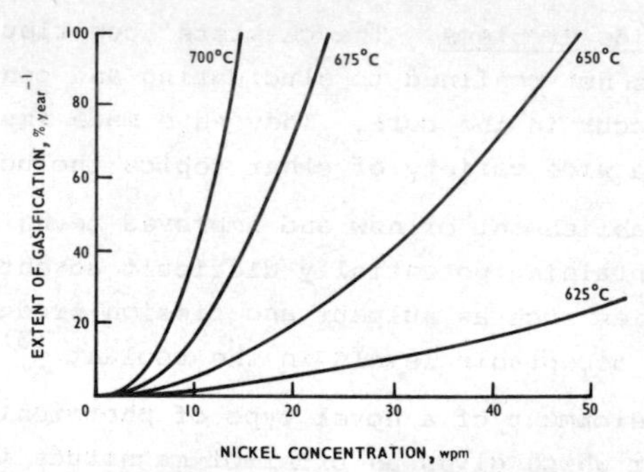

FIG. 10 RELATIONSHIP OF REACTION RATE TO NICKEL CONCENTRATION AND TEMPERATURE FOR THE OXIDATION OF A PURE GRAPHITE IN AGR COOLANT AT AMBIENT PRESSURE

of nickel contamination at four temperatures between 625°C and 700°C.

These results have had a direct impact on reactor operation. For short shutdowns, temperatures are maintained above 250°C; for longer shutdowns and outages, the coolant carbon monoxide concentration is reduced to below 0.2 vol% before the temperature is allowed to fall below this level.

As an extension to this work, the possibility of inhibiting nickel-catalysed oxidation has been investigated. It has been shown that the presence of traces of silicon on the graphite substantially reduces corrosion relative to the nickel-catalysed case. Mechanistic details remain to be settled, but available data suggest that a 20:1 silicon-to-nickel atomic ratio is generally required to achieve substantial protection.[35]

5. Waterside Corrosion and Water Chemistry Control

5.1 Water Purity Control. Power station boilers use large quantities of water (approximately 1 litre sec^{-1} for every megawatt generated, which means about 44,000 l s^{-1} at times of maximum electricity demand). They convert the water to steam and then superheat it; they also act as giant concentrators of any salts which are present.

Although boiler feed-waters are generally very pure, several trace contaminants are usually present, including sodium, magnesium, potassium, ferrous, cupric, nickel, sulphate, chloride, phosphate and acetate ions. As they concentrate in the boiler, these ions behave in a variety of ways; sparingly soluble salts and transition metal oxides precipitate before reaching a high concentration, while the more volatile salts evaporate into the steam; others remain in the concentrating liquid film on the tube surface and, if present in sufficient quantity, can create a serious corrosion risk. It is therefore necessary to set stringent limits on the solute levels which can be tolerated.[36]

The extent to which boilers can generate high local concentrations of salts is demonstrated in Figure 11. This

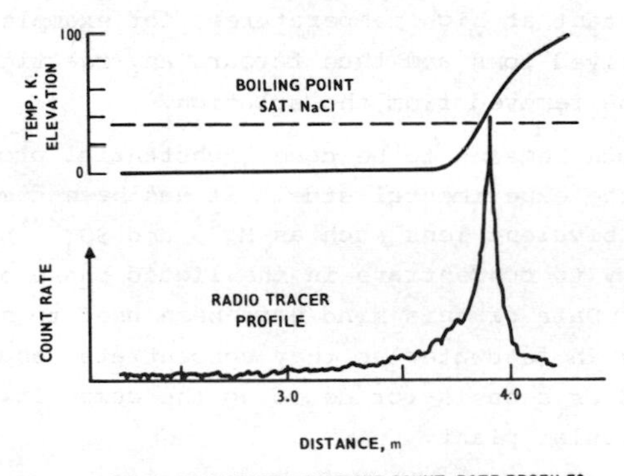

FIG. 11 TUBE TEMPERATURE AND COUNT-RATE PROFILES FOR CONTINUOUS SALT DEPOSITION FROM AN AQUEOUS SODIUM CHLORIDE SOLUTION

shows the results of an experiment in which a solution containing approximately 1 mg kg^{-1} of radio-active sodium chloride ($^{24}NaCl$) was evaporated in a boiler tube. The solution was already boiling when it entered the tube (at 0m on the distance axis), and boiling continued along the tube until, at final dryout, the tube wall temperature rose, a concentrated liquid solution film developed, and solid sodium chloride eventually deposited. The increase in count rate clearly shows the formation of both concentrated solution and deposit. Concentration factors in excess of 10^5 have been shown to develop under such circumstances, with the result that benign dilute boiler water can become locally extremely aggressive.[37]

In quantifying water quality, it is important to understand how each individual species will behave (and influence the behaviour of others) and much careful chemical research has been carried out by the Board's chemists to establish the relevant data.[38] The experimental techniques are extremely exacting, as they involve working with aqueous solutions at pressures of 180 atm and temperatures up to 370°C (the critical point of water).

Chemical assessments are complicated by the influence which water itself has on the behaviour of solute ions at the high temperatures and pressures of interest. Figure 12 shows that the properties of water (compressibility, density, dielectric constant) vary substantially with changes in temperature. This in turn affects the stabilities, solubilities and degrees of solvation of the dissolved ions. The reduction in dielectric constant at high temperatures, for example, destabilises dissolved ions and thus favours any reaction by which they would be removed from the solution.

While much remains to be done, substantial progress has been made on the experimental side. It has been demonstrated, for example, that bivalent ions such as Mg^{2+} and SO_4^{2-} have a particular tendency to concentrate in the liquid phase at liquid-steam boundaries. Data of this kind have been used to model the behaviour of salts in feedwater as they concentrate, and these models have served as a basis for deciding the composition of feedwater in particular plant.

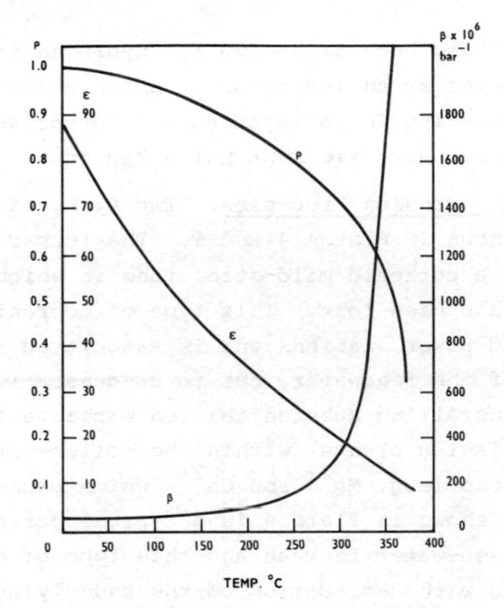

FIG. 12 VARIATIONS IN DENSITY, ρ, DIELECTRIC CONSTANT, ε, AND COMPRESSIBILITY, β, OF WATER WITH TEMPERATURE, ALONG THE SATURATION LINE

Figure 13 sets out the limits which are imposed on the most highly rated units, the once-through AGR boilers. Values

OXYGEN	$\leqslant 5\ \mu g\ kg^{-1}$
CONDUCTIVITY AFTER CATION EXCHANGE	$\leqslant 0.1\ \mu S\ cm^{-1}$
TOTAL Fe + Cu + Ni	$\leqslant 10\ \mu g\ kg^{-1}$
SODIUM	$\leqslant 5\ \mu g\ kg^{-1}$
SILICA	$\leqslant 20\ \mu g\ kg^{-1}$
Cl^- AND SO_4^{2-}	$\leqslant 2\ \mu g\ kg^{-1}$

AMMONIA IS DOSED TO GIVE A pH OF 9.4 AT 25°C
10 TO 20 $\mu g\ kg^{-1}$ OF HYDRAZINE IS ALSO ADDED.

FIG. 13 FEEDWATER TARGETS FOR AGR BOILERS

are included for sodium, silica, transition metals and oxygen, together with the recommended conductivity for the water after all the cations have been exchanged for hydrogen ions. Conductivity serves as an indicator of anion concentration, with silica accounting for a major part. Corrosive anions such as chloride and sulphate are kept below $2 \mu g\ kg^{-1}$.

5.2 Waterside Corrosion Processes. Two types of waterside corrosion are shown in Plates 4 and 5. The former shows a cross-section through a corroded mild-steel tube in which above half the wall thickness has been lost. This type of corrosion occurs on sea-water-cooled power stations and is associated with sea-water contamination of the feedwater, due to condenser leakage. It is controlled by installing substantial ion exchange facilities (condensate polishing plants) within the boiler-water circuits to remove the ions (e.g. Mg^{2+} and Ca^{2+}) which promote it. The laminated oxide shown in Plate 4 is a typical corrosion product following from sea-water ingress and this type of corrosion is also often associated with degradation of the underlying metal by a phenomenon known as hydrogen damage. Hydrogen from the corrosion reaction diffuses into the steel where it attacks the carbides to form methane and other hydrocarbons, thus weakening the metal. Eventually the tube fails catastrophically, a "window" being blown out of the tube wall.

Plate 5 shows an example of stress-corrosion cracking, another type of damage which can be caused by sub-standard boiler water. It results from a combination of electrochemical and mechanical forces acting on the metal. In the example shown, the material is an austenitic stainless steel of the kind used in superheaters, where it may come into contact with concentrated solutions formed by the evaporation of water droplets entrained in the steam. Alkaline concentrates are particularly aggressive in this situation. The 3 mm long crack shown in the left hand side of Plate 5 propagated in only eight hours through a superheater tube specimen exposed to sodium hydroxide.

The susceptibility of a steel to this type of corrosion depends very sensitively on the exact composition of the salt solution. The second crack in Plate 5 is shown at ten times the magnification of the first. The only difference in experimental conditions was the presence of a small proportion of sulphate in the sodium hydroxide in the latter case. Many other ions have similar specific effects and it can be very difficult to determine what cracking risk a particular contaminated boiler feedwater represents. Board chemists are continuing to investigate the

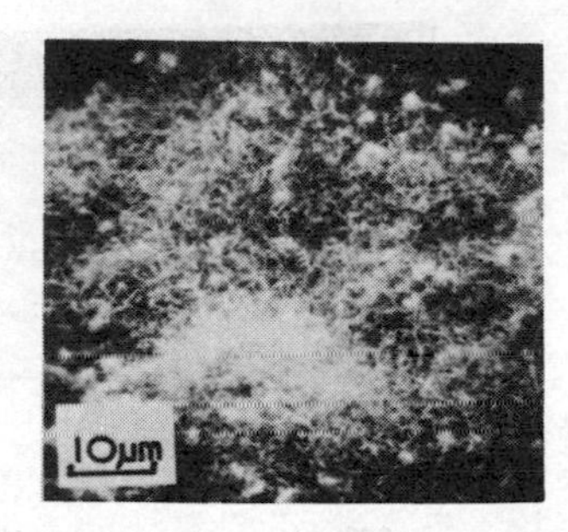

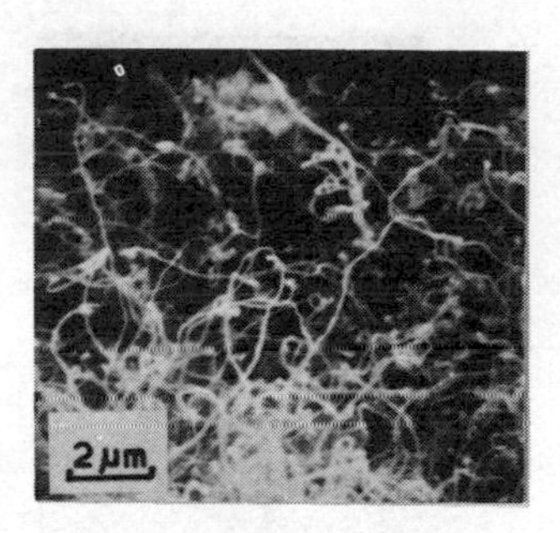

PLATE 3 FILAMENTARY STRUCTURES ON 20/25 Nb STEEL AFTER REACTION IN CO AT 550°C

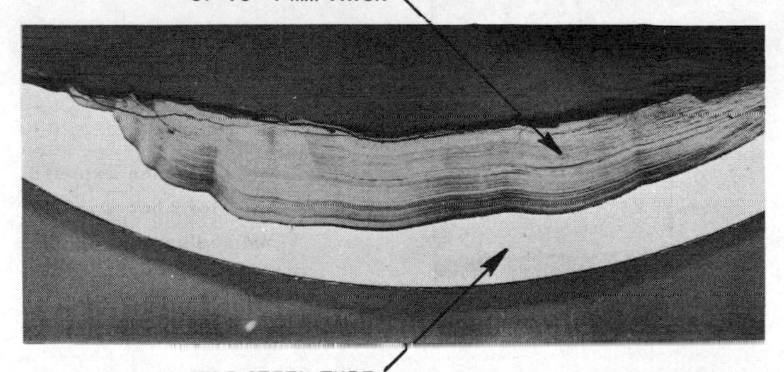

PLATE 4 CROSS SECTION THROUGH LAMINAR SCALES FOLLOWING WATER-SIDE CORROSION AS A RESULT OF SEA-WATER INGRESS

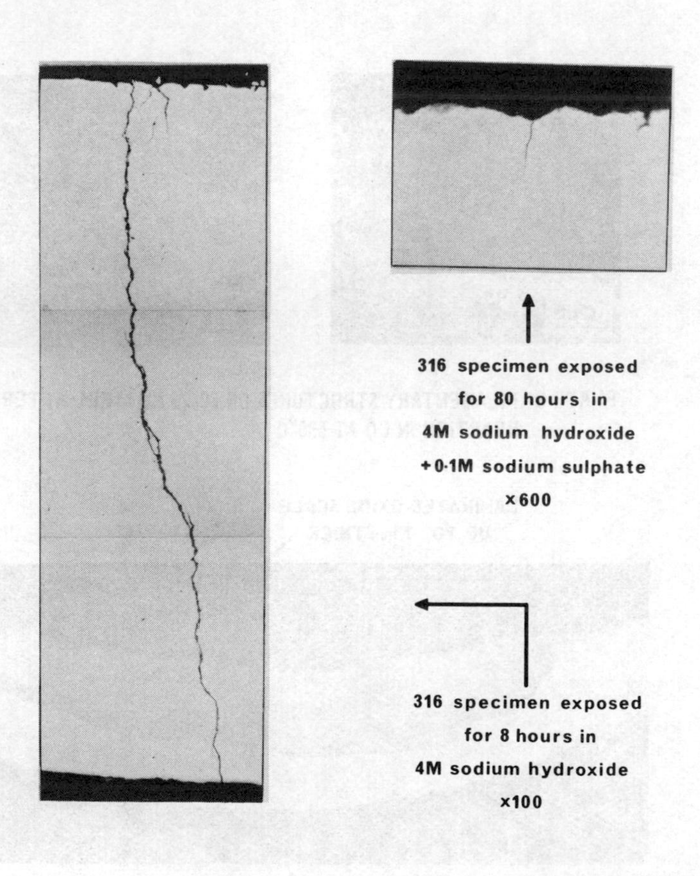

316 specimen exposed
for 80 hours in
4M sodium hydroxide
+0·1M sodium sulphate
×600
316 specimen exposed
for 8 hours in
4M sodium hydroxide
×100

PLATE 5 INHIBITION OF CAUSTIC STRESS CORROSION OF 316 SS BY ADDITION OF SODIUM SULPHATE

mechanisms involved.

5.3 Decontamination of Pressurised Water Reactors. In some types of nuclear power plant, water is used not only as the working fluid of the steam cycle but also as the primary coolant and moderator of the reactor core. The water chemistry of the primary cooling circuit is thus inevitably complicated by the presence of nuclear radiation. CEGB chemists, however, have been able to turn this factor neatly to advantage in solving a potential problem in the pressurised-water reactor (PWR), one of which the Board plans to install at Sizewell.

The interaction of water and high-alloy steel surfaces at high temperature in the coolant circuit produces corrosion products (such as oxides) which tend to become entrained in the flowing coolant and become radioactive under neutron irradiation in the core. The oxides subsequently deposit in parts of the coolant circuits remote from the core (e.g. on bends and in valves) where their radioactivity increases the radiation dose which the maintenance staff could receive.

The problem could be overcome if some means was available for dissolving these oxides into the water stream. Board chemists have now made this a practical proposition through the development of the LOMI (low-oxidation-state metal ion) reagent. The main constituents are a powerful one-electron reducing agent (such as Cr^{II}, V^{II} etc) which dissolves radioactive oxide through the reduction of Fe^{3+} ions in the oxide, a complexing agent which maintains both spent reductant and dissolved metal ions in solution, and a buffer to keep the pH constant. As originally formulated, the reagent was based on vanadous picolinate, with acetate as the buffer.

The reagent was designed to be applied to reactors during their annual overhaul, and to minimize downtime it is advantageous to decontaminate as soon as possible after shutdown, when the gamma radiation field within the reactor core is still high. However, gamma irradiation generates short-lived intermediate species in the water (hydrated electrons, hydroxyl radicals (OH) and hydrogen atoms (H), together with small amounts of hydrogen peroxide) which have a net oxidising effect on the vanadous picolinate and so reduce its effectiveness. Taking this into account, it was estimated that the useful lifetime of reagent under typical reactor decontamination conditions would be only about six hours, suggesting that a substantial fraction would be wasted. It was thus desirable to build in some additional stability.

One way of doing this would be to eliminate some or all of the species responsible for the oxidation process e.g. the H, OH and H_2O_2, the hydrated electron being a powerful reducing agent). The chemists' proposal was to substitute formate for acetate as the buffer. The formate ion scavenges H and OH to yield carboxyl radicals (CO_2^-) which are a reducing agent nearly as reactive as the hydrated electron. In the presence of excess formate it was shown that radiolysis of $V^{II}(pic)_3^-$ does not result in oxidation. There was, in fact, a slight increase in the $V^{II}(pic)_3^-$ concentration as the small amount of V^{III} impurity inevitably present in these solutions was reduced.

$$CO_2^- + V^{III}(pic)_3 \rightarrow CO_2 + V^{II}(pic)_3^-$$

From the plant point of view this is important as it means that the presence of excess formate in the decontaminating reagent not only ensures that H and OH do not bring about the oxidation of $V^{II}(pic)_3^-$, but also that $V^{III}(pic)_3$ formed in the oxide dissolution reaction can be converted back to the starting material[40]. The pattern of reaction is shown in Figure 14.

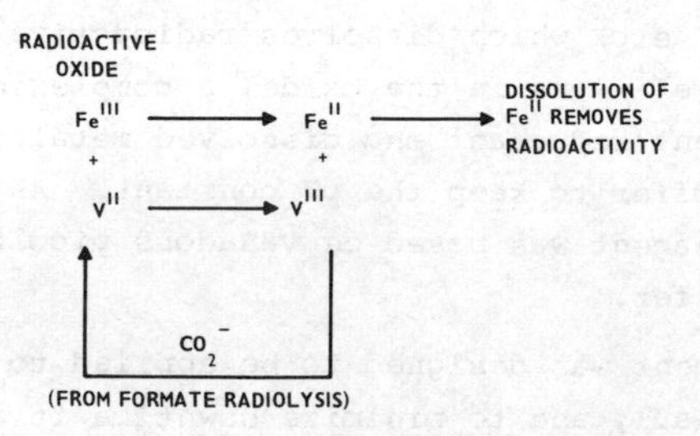

FIG. 14 THE RADIOLYTIC REGENERATION OF VANADIUM II DECONTAMINATION REAGENT

This can be mimicked in the laboratory by successively exposing a $V^{II}(pic)_3^-$ solution to the atmosphere (the V^{II} is oxidised rapidly by oxygen, a process equivalent to reductive dissolution) and then to a flux of γ-rays in a ^{60}Co source. The results of such an experiment are shown in Figure 15. The reagent can be cycled like this many times without adversely affecting its ability to dissolve oxide deposits.

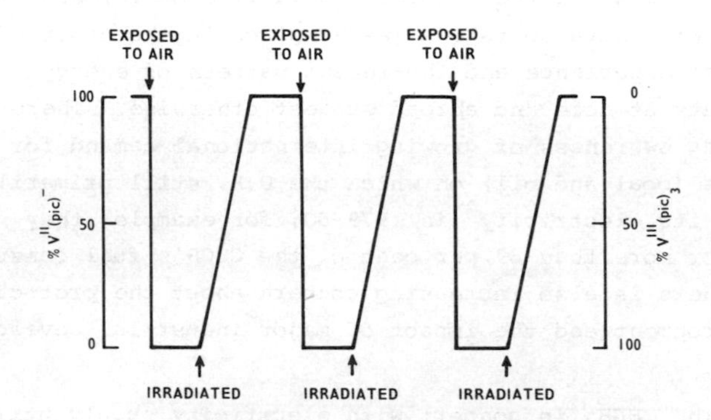

FIG. 15 EXPERIMENTAL DEMONSTRATION OF REGENERATION OF VANADOUS PICOLINATE IN THE PRESENCE OF FORMATE ION

This is believed to be the first time that the radiation field has been actively put to use in the formulation of a decontaminating reagent. Only formate is destroyed in the decontamination process and the product is gaseous CO_2, which is inert and easily removed from the system. The ability of the reagent to regenerate itself means that it is positively advantageous to perform the decontamination as soon as possible after shutdown, when the radiation fields are highest. Designated LOMI reagent, the new formulation has already been used for decontaminating the Winfrith Steam-Generating Heavy Water Reactor.[41]

6. Future Developments

The last three sections give an indication of some major problem areas with which the Board's chemists have been and continue to be deeply involved. It must be stressed, however, that within the electricity supply industry there are many other issues which require the chemist's attention. These extend from measuring fuel qualities to finding ways of disposing of waste products; from developing monitoring techniques and equipment for operational control to identifying improved on-site means of supplying pure water, hydrogen and chlorine (all essential materials for the operation of a modern power station). At the present time, therefore, the chemist plays a vital role in ensuring the efficient operation of the industry.

But what of the future? Can a time be foreseen when the chemist will have to take a lesser place in the order of things? Past experience and the recent pattern of energy related events at home and abroad suggest otherwise. There is an increasing awareness of growing international demand for the fossil fuels (coal and oil) on which the U.K. still primarily depends for its electricity (in 1979-80, for example, they accounted for more than 89 per cent of the CEGB's fuel consumption[22]). There is also increasing concern about the protection of the environment and the impact of major industrial developments upon it.

The CEGB, in concert with electricity supply utilities throughout the world, is giving much thought to ways in which future constraints on fuel supplies and environmental impact might most economically be met. The chemist may be expected to feature prominently in developments, which could result in radical changes in the design of future power stations and the ways power is transmitted to the consumer. A good example in the generation field is the gasifier combined-cycle concept for coal burning power stations.

This type of plant has been touched on in earlier presentations at this meeting, but briefly it involves first gasifying the coal (using air, oxygen or steam, according to the particular properties required of the product fuel gas). The resulting fuel gas is cleaned, to remove dust and environmentally sensitive elements such as sulphur, and is then burned in a gas turbine. The exhaust gases are then passed through a conventional steam-raising boiler.

The attraction of such a complex unit lies in its ability to combine a high degree of environmental control with a better fuel-to-power conversion efficiency (40 to 43%), than that available from conventional coal-fired power stations. The latter would have an overall conversion efficiency of only 35 to 36%, when suitably modified with, for example, flue-gas desulphurisation (fgd) to achieve an equal degree of control. Hence, if at sometime in the future a substantial reduction below present levels was asked for in the sulphur emissions from power stations, then the new plant could achieve this while producing 10% to 20% more electricity than fgd-modified conventional plant, per tonne of coal burned. With current CEGB expenditure on coal of over £2B per annum, the financial attractions are clear; such novel plants would, however, require substantial research and development effort, much of it chemistry based, before they could be exploited on a major industrial scale.

There are further intriguing options. For example, rather than burn the fuel gas directly in an adjacent plant, it might be piped to points of high local load and there burned in a fuel cell to produce not only electricity but also heat, which could be used for district heating.

Other possibilities come from the nuclear side of the industry. Nuclear power is significantly cheaper than that from coal- or oil-fired plant and nuclear stations will account for a steadily increasing proportion of the base-load power generated in future decades. It is reasonable to ask how, when the base-load is fully catered for, the price advantage of nuclear stations might be further exploited to meet some part of the peak demand. Could some of this cheap power be stored at periods of low demand in the daily cycle and consumed at peak times, replacing the more expensive direct coal- or oil-fired generation which would otherwise be needed?

Water storage is one possibility, and the Board is currently building a 1500 MW plant at Dinorwic. This will take advantage of low-cost conventional power as well as nuclear. Storage batteries are a second option. Once developed, they would be particularly suitable for points of high or growing load, where expensive grid-transmission reinforcement might otherwise be needed. Storage batteries are also of interest for those situations where renewable sources of power, such as the wind, are being tapped. They would cater for those occasions on which temporary fluctuations in the source reduce direct supplies below demand levels. Yet another storage alternative would be to use the cheap nuclear power to generate hydrogen by the electrolysis of water. The hydrogen could then be used either to generate power in a fuel cell, or as a basic feedstock for the chemical industry.

It should also be borne in mind that because power stations are extremely expensive to build, there will be an increasing need, as existing plants age, for chemists to help find ways of extending their useful lifetimes, and new types of plant already scheduled for the system (e.g. the pressurised water reactor) will demand the development and application of chemistry based skills. The sodium-cooled fast breeder reactor may bring further issues needing investigation; this is, in fact, an area where the Board is already mounting a small research effort.

The above list is by no means exhaustive, but it is certainly reasonable to conclude from it that the next few decades will be at least as challenging for the electricity supply industry as anything that has been faced in the past. Perhaps one of the few certainties in an uncertain future is that the chemist will continue to play a vital role in charting the path which the industry follows.

7. Acknowledgements

This paper is published by permission of the Central Electricity Generating Board.

8. References

1. J. Priestley, "The History and Present State of Electricity" 3rd Ed., London 1775, p 79.
2. M. Archer, "Electrochemistry Since Davy and Faraday". Proceedings of the Royal Institution of Great Britain", Applied Science Publishers, London, 1976, Vol. 49, p 209.
3. W. Ostwald, "Elektrochemie Geschichte und Lehre", Leipzig, 1896.
4. "Report of the Committee of Inquiry into the Electricity Supply Industry", Her Majesty's Stationery Office, London, 1956. Cmd 9672, p 126.
5. F.A. Williams and C.M. Cawley, Paper 2 in "The Mechanism of Corrosion by Fuel Impurities", ed. by H.R. Johnson and D.J. Littler, Butterworths, London, 1963, p 24.
6. A.B. Hedley, Paper 11 in "The Mechanism of Corrosion by Fuel Impurities", ed. by H.R. Johnson and D.J. Littler, Butterworths, London, 1963, p 204.
7. A.B. Hart and C.J. Lawn, CEGB Research, 1977, No. 5, p. 4.

8. W.D. Halstead and E. Raask, J. Inst. of Fuel, 1969, 42, 344.

9. D. Cubicciotti, High Temperature Science, 1972, 4, 32.

10. A.J.B. Cutler, T. Flatley, and K.A. Hay, CEGB Research, 1978, No. 8, p 12.

11. A.B. Hart, J.W. Laxton, in "The Efficient Uses of Energy", ed. by I.G.C. Dryden, IPC Science and Technol. Press, Guildford, 1975, Chapter 20, p 505.

12. W.D. Halstead and J.W. Laxton, J. Chem. Soc. Farad. Trans. I, 1974, 70, 807.

13. A.W. Coats, D.J.A. Dear and D. Penfold, J. Inst. of Fuel, 1968, 41, 129.

14. W.D. Halstead, J. Inst. of Fuel, 1969, 42, 419.

15. A.S. Kallend, Combustion and Flame, 1967, 11, 81.

16. A.W. Coats, J. Inst. of Fuel, 1969, 42, 75.

17. Year Book of the Royal Society, London, 1980, p 226.

18. J. Bettelheim, W.D. Halstead, D.J. Lees and D. Mortimer, Proceedings of VGB Conference on 'Research in Power Plant Technology', Essen, 21 - 22 May, 1980.

19. A.J.B. Cutler, W.D. Halstead, J.W. Laxton and C.G. Stevens, J. Engineering for Power, 1971, 307.

20. K.G. Saunders, J. Inst. of Energy, 1980 53, 109.

21. J. Bettelheim and W.W. Hann, J. Inst. of Energy, 1980, 53, 103.

22. Central Electricity Generating Board, Annual Reports and Accounts, 1979-80, C.E.G.B. London, p 13.

23. C.J. Wood and A.J. Wickham, Nuclear Energy, 1980, 19, 277.

24. J.V. Shennan, "Gas Chemistry in Nuclear Reactors and Large Industrial Plant", Heyden, London, 1980, p 98.

25. P. Campion, "Gas Chemistry in Nuclear Reactors and Large Industrial Plant", Heyden, London, 1980, p 53.

26. A.J. Wickham, J.V. Best, and C.J. Wood, J. Radn. Phys. Chem., 1977, 10, 107.

27. J.V. Best, "Gas Chemistry in Nuclear Reactors and Large Industrial Plant", Heyden, London, 1980, p 141.

28. D.J. Norfolk, R.F. Skinner and W.J. Williams, J. Radn. Phys. Chem. In press.

29. M.P. Hill, A.M. Brown, Carbon, 1981, In press.

30. A.M. Brown, A.M. Emsley and M.P. Hill, "Gas Chemistry in Nuclear Reactors and Large Industrial Plant", Heyden, London, 1980, p 26.

31. M.J. Bevan, Proceedings of "Carbon '80, Third Int. Conf. on Carbon", Deutschen Keramischen Gesellschaft EV, Baden-Baden, July 1980.

32. M.J. Bevan, and B.H.M. Billinge, Proceedings of "Carbon '80, Third Int. Conf. on Carbon", Deutschen Keramischen Gesellschaft EV, Baden-Baden, July 1980.

33. A.N. Freedman, J. Chromatography, 1980, 190, 263.

34. A.M. Brown, J. Graham, K.G. Saunders and P.L. Surman, Corrosion Science, 1978, 18, 337.

35. S.D. Mellor, Central Electricity Research Laboratories, Leatherhead, to be published.

36. G.M.W. Mann, British Corrosion J., 1977, 12, 6.

37. R. Garnsey, Proc. British Nuclear Energy Society Conf. on Ferritic Steels for Fast Reactor Steam Generators, Paper 64.

38. D.J. Turner, "Thermodynamics of Aqueous Systems with Industrial Applications" ed, S.A. Newman, p 653.

39. D.J. Finnigan, R. Garnsey, D.F. Libaert, P.D. Allen, Proc. British Nuclear Energy Society, Second Int. Conf. on Water Chemistry of Nuclear Reactors", paper 8.

40. R.M. Sellers, J. Radn. Phys. Chem., In press.

41. D. Bradbury, M.G. Segal, R.M. Sellers, T. Swan and C.J. Wood , Water Chemistry II. British Nuclear Energy Society, 1980, p 279.

The Chemistry of Power Station Emissions

By A. B. Hart
CHEMISTRY DIVISION, CENTRAL ELECTRICITY RESEARCH LABORATORIES, KELVIN AVENUE, LEATHERHEAD, SURREY KT22 7SE, U.K.

Introduction

A modern coal- or oil-fired power station is a highly concentrated source of combustion gases, and rather special measures have to be taken to protect the environment from pollution by them. These comprise a combination of control of the unwanted emission at source, for example by removal of dusts, and by dispersal of the bulk of the flue gases from very tall chimneys. The effectiveness of tall stacks has been argued in detail by Clarke, Lucas and Ross[1]. A 2000 Megawatt pulverised coal-fired power station chimney, or rather set of chimneys, because there will be four of them enclosed in one chimney stack (see Figure 1), will discharge nearly 2000 tons of carbon dioxide for every hour it is working at full load, i.e. about 12% of the dry flue gas volume.

Figure 1
Flue gases emerging from the 200 m high chimney of Fawley power station which is situated on the South bank of Southampton water. It has four 500 MW turbines powered by oil-fired boilers.

Carbon dioxide is the least regarded as a pollutant, although since the early 70's the possible long-term climatic effect, on a global scale, of adding to the long-established atmospheric concentration of carbon dioxide has become a subject of speculation and enquiry[2].

The principle reason is that the measured atmospheric concentration of CO_2 at different latitudes has increased by about 7% over the last 25 years and the underlying trend matches the trend in global CO_2 emissions from fuel burning. About half of the CO_2 emitted appears to have remained in the atmosphere during the available period of observation, the rest being absorbed by vegetation, soil and the oceans. A primary effect of increasing the CO_2 concentration is to increase the absorption of infra-red radiation from the earth's surface and so lead to a warming of the atmosphere (the so-called "greenhouse" effect). Models used to predict future energy demand have been used in predicting future atmospheric CO_2 concentrations and lead to estimates of further increases of up to 20% over the next 25 years. Concern has therefore been expressed that a 2% - 4% annual growth in the total amount of fossil fuel burned which was expected before the current world recession in industrial activity could lead to significant climatic effects in the future. If the atmospheric CO_2 content continues the linear increase shown over the past 25 years then the CO_2 concentration will take about 350 years to double in value and it has been estimated that this would raise surface temperatures by about 2°C, but with the above annual growth scenario this stage would be reached around 2025 AD. Such a temperature rise would certainly cause important climatic changes with massive consequences such as flooding of coastal areas and major cities.

Along with the carbon dioxide as it emerges from the stack there will be three hundred tons of water vapour and rather more than half a ton of very fine dust carried along with the gases which will emerge from the chimney top at the high velocity of 30 m/s. This dust consists of spherical micro-particles of diameter 10 μm or less of aluminosilicates plus some particles of silica associated with traces of unburnt coal and soot. Adhering to these particles are minute amounts of inorganic salts and oxides condensed from vapours driven out of the coal minerals during their passage through the furnace. Sodium sulphate predominates, oxides or sulphates of iron, calcium and magnesium are present in lesser amounts, and there are minute but detectable traces of compounds of elements derived from the coal such as lead, cadmium, antimony, arsenic, chromium, manganese and vanadium. Table 1 lists some 'impurities' in coal.

Table 1 - Coal Impurities

'Impurities' in a typical coal (parts per million by weight)			
Sulphur	16000	Zinc	60
Nitrogen	12000	Phosphorus	1000
Silicon	26000	Chromium	60
Vanadium	145	Cobalt	47
Iron	13200	Manganese	85
Nickel	25	Copper	130
Calcium	9500	Lead	50
Potassium	2300	Selenium	7
Aluminium	25500	Cadmium	3
Sodium	1470	Antimony	2
Chlorine	3400	Arsenic	10
Magnesium	3700	Mercury	0.3

Table 2 gives, for comparison, the impurities in a typical heavy residual fuel oil. With the exception of sulphur the impurities are much less. This paper deals mainly with coal-firing which now accounts for most of the electricity generated from fossil fuels in the U.K.

Table 2 - Heavy fuel oil impurities

'Impurities' in oil (parts per million by weight)			
Sulphur	20000	Zinc	4
Nitrogen	1500	Phosphorus	4
Silicon	300	Chromium	3
Vanadium	150	Cobalt	3
Iron	100	Manganese	2.5
Nickel	50	Copper	2.5
Calcium	100	Lead	2
Potassium	50	Selenium	1
Aluminium	75	Cadmium	0.2
Sodium	50	Antimony	0.02
Chlorine	25	Arsenic	0.01
Magnesium	12	Mercury	0.01

Sulphur is an important impurity - 1.6 percent in the average coal (with a range of 1.0 to 2.5 percent) and 2.0 percent in heavy fuel oil. It occurs both as organic sulphur compounds in the coal substance and as inclusions of pyrites (iron sulphide, FeS_2). It burns to sulphur dioxide and up to one percent of this may oxidise in the hot gases to sulphur trioxide which eventually unites with the water to form sulphuric acid which mostly reacts with alkaline oxides in the ash.

Sulphur dioxide is the chief obnoxious component of flue gases. In an hour's operation at full-load there will be 28 tons of sulphur dioxide released from the chimney at a rate of 1.3ℓ of the gas for every m^3 of flue gas. Nitrogen is another important impurity in coal occurring as amino, imino and ring nitrogen groups. Much of it is oxidised to nitric oxide, NO. At the very hottest parts of a flame nitrogen from the air reacts with oxygen to form the equilibrium amount of nitric oxide appropriate to that temperature. From these two sources - nitrogen from the air and fuel nitrogen, the flue gases acquire between 0.3 and 0.6ℓ of NO per m^3 and the 2000 MW power station would expel into the atmosphere 3 to 6 tons of NO per hour at full-load. A small but highly variable fraction of the NO is oxidised to the dioxide, NO_2/N_2O_4 which is the physiologically active form, NO being itself inert. This oxidation is thermodynamically favoured when the flue gases mix with air though it occurs extremely slowly. It has become customary to refer to nitrogen oxides as NO_x since NO is capable of slow oxidation and can be regarded as a potential source of the more active oxide.

Another gaseous impurity in the flue gas is hydrochloric acid. Derived from the average 0.24 percent of chlorine in coal this will amount to 2.5 tons of the acid gas per hour. Carbon monoxide is also present (1ℓ per m^3) in the flue gas - about 9 tons/hr from the 2000 MW power station.

It is not difficult to calculate the height at which a plume must be discharged to ensure that these gases have been adequately diluted before the dispersed plume touches ground level (e.g. see Clarke, Lucas and Ross 1970[1]). The dilution required is decided for SO_2, the most plentiful potential pollutant, but applies to other gases and also to particles of size so low (< 10μm diameter) that they behave more or less as gases in respect of diffusion. Larger particles of ash which would not be subject to the same dilution because they would fall out under gravity have to be removed from the flue gases before discharge. With a coal-fired power station this is normally achieved electrostatically. The dust-laden gases are passed between a wire cathode and an earthed anode with 40 to 50 kV between them. For dust particles up to about 20 μm in diameter the electrical and drag forces combine to give a deposition velocity which increases roughly in direct proportion to particle size to a value giving a collection efficiency very close to 100 percent, the deficiency being due to re-entrainment which is negligible if the particles have an electrically conducting film on their surface as they normally do. The electrostatic precipitators are designed to limit emitted dust to 0.115 mg/m^3 which is the value currently set by the Alkali Inspectorate for modern power stations and is related to the physical rather than the chemical properties of the dusts. Figure 2 illustrates the features of an electrostatic precipitator.

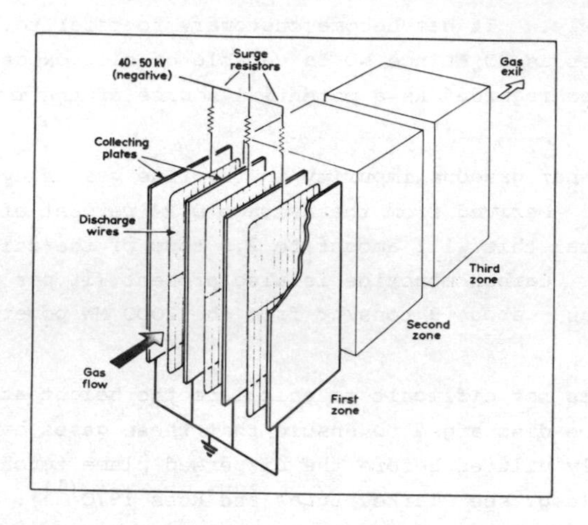

Figure 2. Schematic arrangement of wire grids and plates in an electrostatic precipitator used with coal-fired boiler plant. Solid particles become electrically charged in the high field region near the wires and are then attracted to the collecting plates. The accumulated layer is dislodged by mechanical rapping. Current designs of precipitators are intended to comply with a dust emission limit of 115 mg/m^3.

As we shall note later the smaller particles of dust which make up the fraction escaping the precipitators are likely to be enriched in those components of the flue gases which deposit from the vapour phase on to available surfaces as the gases cool from the original flame temperature. This includes most of the 'trace' metals mentioned above and also alkali sulphates and, indeed, sulphuric acid. This enrichment is due simply to the fact that weight for weight these small particles provide the greater surface area.

Plume dispersal from a tall power station chimney stack begins with a high efflux velocity and thermal buoyancy due to the fact that the gases are normally at a temperature of 120°C or more. The result is that the plume rises to a height over 1½ times that of the stack, the precise height depending on the velocity of the wind which then carries the plume along. The behaviour of the plume depends on air turbulence but generally speaking parts of the plume will not reach the ground from a typical stack of 200 m high until 5 to 10 km from the stack. By that time the aerial turbulence and mixing will have diluted the original plume by a factor of 10^4. This first stage of dispersal takes about an hour and chemical transformations in the plume are not marked during this time. What happens to the plume components thereafter does depend considerably on chemical reactions; for example, on the transformation of sulphur dioxide to sulphuric acid and its reaction with ammonia already in the atmosphere and with nitrogen oxides. Figure 3 illustrates the various stages in chimney plume dispersal.

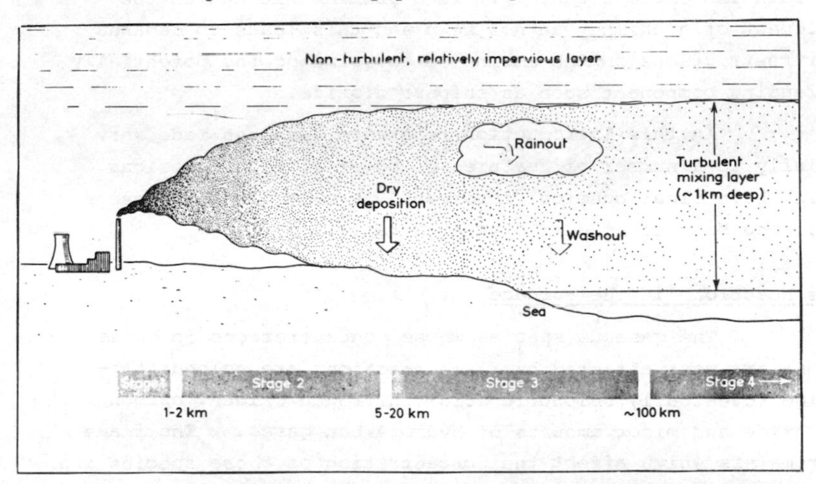

Figure 3. The four main stages of chimney plume dispersion. Ground level concentrations of SO_2, NO_x and fine dust emission are first detected 5 - 10 km from the stack when the plume first descends to the ground. At this point the plume will have been diluted by a factor of 10^4.

Beyond about 20 km the remnants of the original plume begin to be dispersed through the bottom layer of the atmosphere, known as the turbulent mixing layer beneath the so-called thermal inversion boundary. By about 100 km the plume gases will be uniformly dispersed in this layer. Figure 3 also defines the areas of interest in the chemistry of power station emissions. First there are the reactions within the power station which give rise to the emissions in the form in which they emerge. Then there is the near-field, i.e. distances up to, say, 50 km from the power station, where the maximum ground level effect of potential pollutants occurs. Finally there is the far-field, i.e. > 50 km from the station where interest in recent years has centred, particularly in Europe, on the long-range atmospheric transport of acid from coal-burning, heavily industrialised regions like the U.K., Poland and Germany across national frontiers to, for example, Scandinavia. It is maintained that the acid-forming gases persist for over 1000 km from source and elevate the acidity of rain in remote areas. It is a problem with which the tallness of a chimney cannot help and this leads to demands for restrictions on the amount of emission of the potentially offending component such as sulphur dioxide.

In this introduction reference has been made very briefly to a number of features of power station emissions which have clear chemical interest. A selection of these will now be discussed in greater detail.

Gas Reactions in the Furnace

The gaseous species whose concentrations in emissions are most affected by flame reactions are sulphur trioxide (eventually sulphuric acid), nitrogen oxides, carbon monoxide and minor amounts of hydrocarbon gases. The flame parameters which affect the concentration of these species are the temperature to which gas mixtures are exposed and the duration of such exposure, the gas mixing, i.e. the excess oxygen concentration during the combustion process and finally the rate of cooling. In a flame of a particulate fuel (pulverised coal particles or atomised oil droplets) and air the combustion parameters will vary across the width of the reaction zone (see Figure 4).

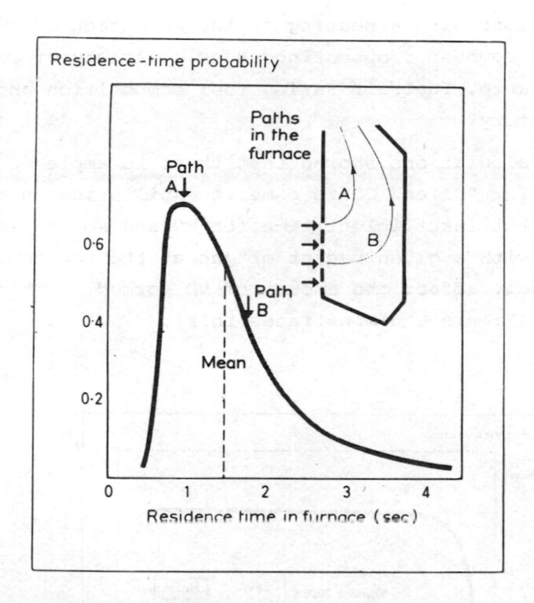

Figure 4. Diagram showing the probability of different portions of flue gas in a power station boiler experiencing different residence times. The inset suggests that the gases originating from one burner will experience a different thermal history from those from another burner.

Some regions will be fuel-rich, heated less by reaction than by radiation and convection, and pyrolysis of the fuel may occur more than in other zones of the flame where oxygen will be in excess. Such different zones will be mixed by violent short-range turbulence. The task of accurately modelling such a complex process in which chemical reactions and mixing are closely involved is a daunting one but considerable success has been achieved with simplified models which assume that at the higher temperatures in a flame most of the chemical species are in dynamic equilibrium with each other and that it is valid to apply known data for the kinetic parameters of the reaction of interest. This approach has made it possible to deduce the effect of cooling and mixing processes on the final gas composition. Thus, Billingsley, Kallend and Marsh[3] took account of the interaction of 46 gas phase reactions with a bearing on the formation of NO, SO_3 and CO in a combustor operating under a variety of conditions of pressure, fuel/air ratio, fuel composition and temperature history.

The calculations showed that there is ample residence time for SO_3 and CO to come to equilibrium in a boiler furnace but insufficient time for NO and so the actual residence time with a given packet of gas at the top temperature will markedly affect the amount of NO formed. This is not so for SO_3 (Figure 5 demonstrates this).

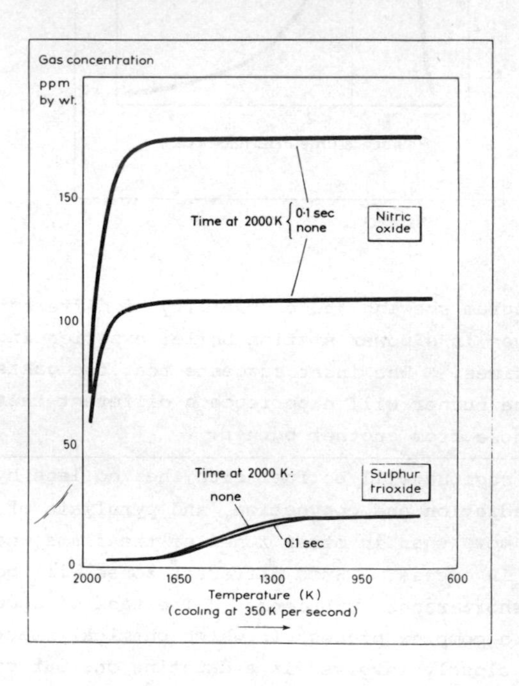

Figure 5. The calculated effect of different residence times at the top temperature in a flame on SO_2 and NO concentration in the cooling gases.

The calculations show clearly that the way to limit the formation of NO is so to arrange the mixing of the gases that some part-cooled combustion gas is recycled to dilute the flame gases and diminish the top temperature achieved. This technique has been applied successfully to boiler plant in the U.S.A. It is also clear from the calculations that SO_3 can be minimised by limiting the concentration of residual oxygen present after the flame and during the cooling of the flue gases from, say, 1650 K to 1000 K. With a pulverised coal flame too low an excess oxygen causes an unacceptable proportion of unburnt carbon and an excessive CO content in the exhaust - both causing a loss of thermal efficiency. A compromise is usually struck with oxygen excess with CO at about 500 - 1000 ppm and SO_3 at 10 ppm. This amount of SO_3 is in fact not unwelcome. With less SO_3 (or H_2SO_4) the electrostatic precipitators often suffer a fall-off in efficiency. Efficient performance of the precipitators can be correlated with the presence of a layer of adsorbed sulphuric acid or alkali sulphate on the fine silicate particles. In the absence of the adsorbed sulphuric acid the smaller particles of dust are not captured on the plates so readily.

Mineral Particles From Burner to Stack

Silicate and aluminosilicate or clay-like minerals, together with silica itself (quartz) amount to 80% of the impurity in coal. A typical thermal history of particles in pulverised fuel in a boiler flame and afterwards when the flue gases in which they are entrained are losing heat to the boiler tubes is as follows (see also Figure 4).

i) a period of rapid heating, mainly by radiation, attaining, in 0.2 to 0.5 s, a maximum temperature of between 1700 and 2000 K depending on the size of the particle;

ii) a cooling period in which the average particle reaches 1000 K in 5 s and 500 K in 10 s as the flue gases approach the chimney stack.

Some at least of the smaller particles, e.g. $< 10\ \mu m$ in diameter or less, which are of interest as possible dust emitted from the stack top will be raised to the higher temperature, e.g. 2000 K. During the short heating period, hydrated silicate minerals undergo endothermic dehydration which begins with the loss of surface adsorbed or inter-lattice water. Fusion of the dehydrated silicate begins as the temperature reaches 1500 - 1600 K depending on the particular mineral. Most fused particles are sufficiently fluid to attain a spherical form by the action of surface forces before cooling begins. Quartz particles except perhaps the very smallest do not fuse although the sharpest edges of crystals are blunted at about 1700 K. The non-silicate portion of the mineral is mainly iron pyrites and carbonates. Pyrites is an impure form of FeS_2 and is associated with trace elements such as lead, arsenic, cadmium, nickel and vanadium. Table 3 summarises the changes which take place in the coal mineral.

Table 3. The Effect of Flame Temperatures on Mineral Species in a Typical Coal

Mineral Species	Weight Per Cent of the Coal Mineral	Changes in the Flame	Products
Potassium alumino-silicates	40	Fusion and vitrification; release of potassium when sodium is present in the gas	Glassy spheres; potassium sulphate
Kaolin	25	Fusion, partial vitrification and re-crystallization	Glassy spheres; mullite, quartz
Quartz	15	Partial fusion and vitrification	Very small amorphous silica spheres
Pyrites	10	Decomposition and oxidation	Magnetite, haematite and SO_2
Carbonates	5	Decomposition and sulphation	Sulphates and CO_2
Chlorides	2	Volatilization and sulphation	Sulphates and HCl

These particles decompose and finish as spherical iron oxide particles (Fe_2O_3 and Fe_3O_4). The more volatile components will be driven out to the gas phase as metal or metal oxide at the flame temperature. Sulphur will be oxidised to the dioxide either during oxidative attack on the pyrites or immediately following its thermal decomposition. Carbonates decompose at about 1000 to 1100 K leaving calcium and magnesium oxides which form sulphates with gas phase SO_2 and O_2 in the cooling flame gases or unite with the silicates and contribute to their latent alkalinity and react with acid sulphate later on. Sodium will appear in the flame gas as NaOH mainly from the coal structure where it is usually present in association with, though not necessarily entirely combined directly with, chloride. Potassium ions escape from some of the minerals, e.g. from muscovite and illite which usually make up 30 - 50% of the mineral matter; muscovite contains 11% of potassium as K_2O, and illite 5 - 6%. The escape of potassium from the hot minerals depends on an exchange with the sodium in the hot gases.

From the point of view of the properties of the emitted ash special interest attaches to the condensation of the volatile species. This takes place on nuclei and available silicate particle surfaces provide suitable condensation nuclei. The result is that the smooth surfaces of the spheres become roughened by tiny excrescences (see Figure 6). The bulk of the condensed material is sodium, potassium and iron sulphates with sodium predominating plus the trace elements referred to above. The finer ash, i.e. $< 10\mu m$ diameter, with higher specific surface area on leaving the stack with the flue gases carries a greater burden of the trace elements or, indeed, of all condensed material than the larger, e.g. $> 100\mu m$ diameter, particulates which are efficiently retained in the electrostatic precipitators. The surfaces of the particles approach approximate equilibrium with the flue gas in the flame and in the stack itself sulphuric acid can begin to condense at its 'dewpoint' of 140 - 150°C. (The sulphuric acid in a coal-fired flue gas is usually in the range of 1 to 5 ppm by volume (approximately 5 to 25 mg/m^3) and the concentration of any condensed acid on a particle will be up to 85% H_2SO_4 on leaving the precipitators.)

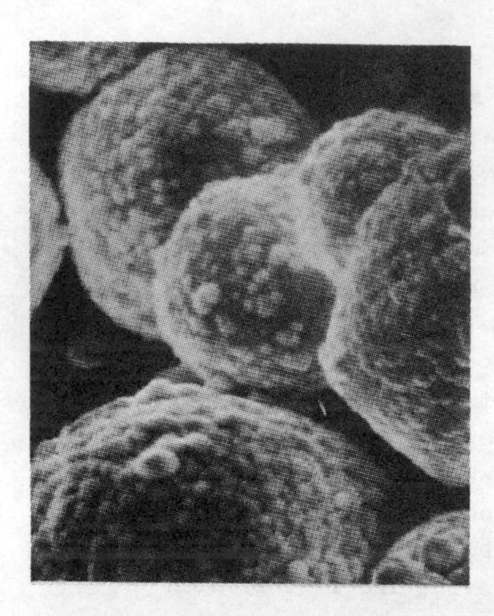

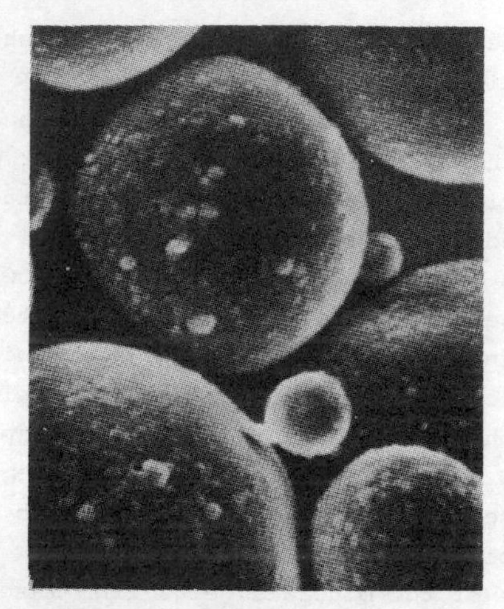

(a) (b)

Figure 6. Fine particles (∿ 1μm in diameter) collected from a coal-fired chimney stack; (a) as collected showing encrustation of deposited sulphates and oxides, and (b) similar particles after washing with dilute hydrochloric acid.

In a typical coal-fired flue gas bearing the maximum permitted amount of 115 mg/m^3 of particulate material mostly in the 2 - 10 μm diameter size range the sulphuric acid absorbed (1.6% S in the coal) would amount to 3 x 10^{-4} to 1 x 10^{-3} g H_2SO_4 per g of ash. This sulphuric acid will react further in two ways - by absorbing moisture from the cooling glue gas and by reacting with the underlying silicate; data collected from typical power stations suggest that neutralisation of the condensed acid by underlying silicate will be complete within the time required for the particles to reach ground level. With a wind velocity of 7 m/s particles would take 14 minutes to reach ground level at a distance 6 km from the 200 m stack (400 m plume rise).

Bulk pulverised coal ash is, of course, widely used as a land-filler (in exhausted quarries and in road works) in concretes and grouts and in the manufacture of lightweight building blocks and aggregates (Raask and Bhaskar[4]). Its properties from the point of view of health or other environmental hazards have been examined with great care. The spherical or rounded particles of fused silicate themselves are quite inert, but the condensed material upon them containing the trace elements and acid is of interest and has been studied by examining the water leachate. Brown et al 1976[5] demonstrated that for all practical purposes the leachable amounts of potentially toxic elements, e.g. arsenic, boron, copper, molybdenum, are below the limits of health hazard. The process of leaching is an interesting one. A fresh ash rapidly releases acid but this is soon neutralised by underlying alkali in the ash and eventually with excess ash in a water sample a pH of 8 - 10 is achieved. Trace elements are released only slowly suggesting that as they have been condensed on the silicate when it is hot some diffusion into the silicate lattice has taken place. Another possible source of health hazard would be the residue of quartz particles, which could consist of between 2 and 15% of the total ash. It has long been recognized that inhaled dusts containing quartz powder can be fibrogenic in man but this property is much diminished if the quartz has been heated above 800°C for 1 h. It has been said that this is due to loss of silanol groups at the surface of the particles. It is known also that Al_2O_3 and Fe_2O_3 can suppress the fibrogenicity of quartz. Both processes may be at work in a coal furnace, for p.f.a. unlike coal mineral, displays no fibrogenic property in standard tests (Raask and Schilling[6]).

The flue-gas-entrained ash particles, amounting to not more than 0.115 g/m^3 of flue gas, constitute only about 0.5 to 0.7% of the total ash but, because of the enrichment mentioned above, 1 to 2% of the total volatile elements will be in the entrained ash. In the air at ground level at the point of maximum concentration this will be diluted to 25 ng/m^3. This refers to all the 'trace' volatile elements - arsenic, cadmium, lead, zinc, copper, chromium. In the U.K. the 'background' concentration in air of these elements is several times this level.

Trace Elements in the Furnace and in the Fuel

The term 'trace elements' connotes those which are present in the coal in parts per million but are of interest because of well-known toxicity of the elements or their compounds - to plants or animals - when available in sufficient concentrations. Understanding the behaviour of these elements is important because of the opportunities which exist for segregating and concentrating the potentially toxic traces. From Table 1 we may select the well-known potentially toxic elements arsenic, lead, cadmium, chromium, nickel, cobalt, copper, selenium and mercury as worthy of consideration, the last being included despite its very low concentration in the coal because it can be expected all to emerge in the gas phase from the chimney whereas the others might be at least partly retained in the ash. These elements are all associated with the sulphidic mineral in the coal and are involved in the exothermic decomposition of the sulphides in combustion.

At 1800 to 2000 K these elements will evaporate, in contrast to the main ingredients of the lithophilic or mainly silicate minerals which fuse but retain much of their minor components. Table 4 lists the chemicals for which these elements are expected to assume in flue gas as the gases cool to the condensation temperature or dewpoint. The data are taken from thermodynamic sources and assume that the condensed phase is pure.

Most of the elements considered would be expected to condense at temperatures in the range 1000 K to 700 K. This is important, because with an aluminosilicate softening point of > 1200 K it means that condensation will be on surfaces which are already hard, hence limiting the degree to which subsequent interactions between condensate and surface can take place and so enhancing their potential mobility once exposed to the environment. It should also be noted that, while for several of the elements listed in Table 4 oxides are the preferred condensed species at the dewpoint, further cooling within the flue gas containing SO_2/O_2 will lead to conversion to sulphates. Estimated conversion temperatures are listed in Table 5.

Table 4 - Preferred Species and Dewpoints of Trace Elements in Coal-Fired Boiler Flue Gases

Element	Concentration Range in Coal ppm	Preferred Species in Vapour Phase	Preferred Species in Condensed Phase	Dewpoint Range K
Antimony	1 to 10	Sb_4O_6	Sb_2O_5	890-910
Arsenic	2 to 100	As_4O_6	As_2O_5	810-850
Cadmium	0.1 to 5	$CdCl_2$	$CdSO_4$	770-830
Chromium	1 to 100	CrO_2Cl_2	$Cr_2(SO_4)_3$	770-840
Copper	10 to 50	Cu_3Cl_3	$CuO/CuSO_4$	800-880
Lead	10 to 100	$PbCl_2$	$PbSO_4$	930-980
Mercury	0.01 to 10	$HgCl_2$	$HgSO_4$	290-350
Nickel	1 to 100	$NiCl_2$	NiO	1250-1890
Selenium	1 to 10	$SeCl_2$	SeO_2	600-660

Table 5 - Oxide to Sulphate Conversion Temperatures in Coal Generated Flue Gases

Oxide	Sulphate	Conversion Temperature K
Sb_2O_5	$Sb_2(SO_4)_3$	450
BeO	$BeSO_4$	800
NiO	$NiSO_4$	1000
CuO	$CuSO_4$	880
SnO_2	$Sn(SO_4)_2$	500

It may thus be concluded that most trace elements will enter the environment in sulphatic form; a notable exception is arsenic where the oxide is preferred.

The Environmental Impact of 'Trace' Elements

The form of the potentially toxic trace elements on leaving the chimney is thus predicted to be a sulphatic deposit on the 1 to 10 μm diameter silicate particles. This prediction is substantiated by measurements on samples collected at chimney top and in the atmosphere. The expected increase in the ratio of furnace-volatile trace metals to ash in the fine 1 to 10 μm stack dust has also been substantiated by many measurements (e.g. Davison et al 1974)[7]. Calculations of the dispersal of stack top ash lead to a maximum three month ground level concentration of the potentially toxic trace elements in the range of 0.05 to 1.0 ng/m^3 which is generally of the order of a tenth to a quarter of the steady background level of such elements in the U.K. Careful measurements of the potentially trace elements have been made in urban neighbourhoods and at sites remote from heavy industrial activity (e.g. in the Lake District). These show that while the steady background concentration of, e.g., arsenic in the rural area is only a half to a quarter that in the comparatively urban region a large power station with a 200 m stack could not contribute even as much of the elements as found in the rural air. The same is true of the deposition of the dust - which follows the same diffusion general laws as deposition of, for example, sulphur dioxide. In the neighbourhood of a coal-fired, 2000 MW, power station an average maximum deposition rate of total particulate matter attributable to the power station has been shown to be of the order 20 mg/m^2/day when the station is running on full-load. But the background deposition at the same semi-rural site was measured as 67 mg/m^2/day. Typical figures for dust deposition in open country have been reported as 15 to 100 mg/m^2/day[8]. Deposition is important because of the effect on soil and plants and particularly because uptake by food plants provides a route to man. This applies to all dust deposition, of course, and the contribution of power station dust from tall stacks does not stand out from normal background.

Smokes and Tarry Organic Materials

By regulation the smoke permitted from a power station stack is severely limited. The total permitted solids can only be achieved if plant conditions - burner efficiency, excess oxygen, gas mixing in the furnace - are such as to consume the fuel particles and avoid the formation of soot and tars almost entirely. The 'unburnt carbon' particles are usually greater than 10 μm in diameter and do not emerge from the chimney. Smaller particles which may be associated with the fine dust do not usually exceed 20 mg/m^3 when the plant is on steady load. At ground level this is a negligible contribution to background smoke and soot. Attention has been focussed upon it, however, because of the well-known presence in such soot particles of the carcinogenic polycyclic aromatic hydrocarbons of which the most dangerous by far is benzo-a-pyrene. The levels of the compound found in the flue gases of modern pulverised coal-fired boilers range between 10 and 400 ng m^3[(9)]. For a 24 h sampling period at the point of maximum ground level impact this would give a concentration range of between 3×10^{-4} and 1×10^{-2} ng m^{-3}. (This range may be compared with a level of 100 ng m^{-3} reported for cigarette smoke.)

Cigarettes and power stations are not, of course, the only sources of this type of compound. Wherever combustion of fossil fuels takes place (e.g. domestic heating systems, garden or municipal refuse fires) formation and release of these compounds will occur, to a degree which reflects the efficiency of the combustion process. Considering benzo-a-pyrene once again modern coal-fired power stations produce much less than 1% of the total entering the environment each year. In effect, these power stations represent the most efficient means of combustion available, a particularly striking demonstration of this being that an open domestic fire yields 10^3 to 10^4 times more benzo-a-pyrene per unit of heat input than does a large boiler.[(10)] The total amount of benzo-a-pyrene introduced into the atmosphere has been the subject of study in the U.S.A. Allen[(11)] concluded that power stations expelled a total of 0.8 tonnes/year whereas residential heating provided 390 tonnes and garden refuse burning 75 tonnes. It is clear that good flame mixing and high intensity combustion in a power station furnace are effective in destroying complex aromatic structures and minimising benzo-a-pyrenes.

Long Range Transport of Power Station Flue Gases

During the last ten years the possibility that industrial emissions could be transported over long distances has become the focus of great interest both in Europe and North America. A central feature of the claims that have been made is that precipitation (rain and snow) is acidic, that there is a trend of increasing acidity and an expanding area of acidity on a regional scale which matches the increased emissions of SO_2 and nitrogen oxides from industry. Such claims are of major concern to electrical power utilities which, through combustion of fossil fuels, are major emitters of both SO_2 and NO_x. In the United Kingdom, for example, electricity generation accounts for about half of the total emissions of SO_2 and nearly half of the NO_x. Sophisticated methods have been successfully developed to protect the near-field environment from these pollutants by ensuring that the flue gases are sufficiently diluted by the air and it has been argued that such measures are likely to be equally effective in removing risk of pollution from the far field. Industry in both Europe and the U.S.A. is, therefore, concerned that it may now be faced with legislation leading to costly and unnecessary additional costs to limit emissions of these pollutants at source.

The initial movement of the plume from a power plant has already been explained; plume constituents are transported in the direction of air-mass flow and (in neutral or unstable conditions) are mixed by turbulent diffusion to fill the atmospheric boundary or mixing layer within 10 - 20 km from the source (see Figure 3). The mixing layer is typically a kilometre or so in depth and is often capped by a temperature inversion and a sharp discontinuity in turbulence. At other times, however, the layer is less well defined and one of the uncertainties in predicting long range transport concerns the validity of assuming that there is a definite mixing height and extent to which material can escape above the mixing layer. Another difficulty in predicting dispersion at long range is that of assessing the relative importance of small scale turbulence and large scale changes in wind direction.

Detailed measurements of windfield and turbulence in an actual plume over long distances coupled with plume traverses at different levels would help to resolve this problem.

Within the mixing layer primary pollutants can be removed by direct contact with land or sea surfaces, which is called 'dry deposition', or by being washed out by rain or snow, i.e. by precipitation scavenging. They can also undergo chemical transformation into secondary products which can then be removed by the same processes.

In the case of SO_2 the dominant removal process is direct dry deposition of the gas but oxidation to sulphuric acid and subsequent deposition as particles of the acid also contributes. Oxidation involves a complex sequence of gas phase reactions in which free radicals, generated photochemically by sunlight, play an important role. Many of the primary reactions occurring have been characterised in detail in laboratory kinetic studies so that it is now possible to model in some detail the chemical evolution of a plume, under conditions where photochemical reactions predominate. The principal chemical step is the reaction of SO_2 with hydroxyl radicals:

$$OH + SO_2 \rightarrow HSO_3$$

leading ultimately to sulphate aerosol. The major reactions generating hydroxyl are

$$O_3 \xrightarrow{h\nu} O\tfrac{3}{4}D + O_2$$

$$O(D) + H_2O \longrightarrow OH + OH$$

$$HNO_2 \xrightarrow{h\nu} OH + NO$$

$$O + RH \longrightarrow OH + R$$

from which it is evident that the fate of SO_2 is connected with other minor atmospheric constituents such as ozone, hydrocarbons and oxides of nitrogen and it is inappropriate to attempt to describe the fate of a single pollant in isolation. Detailed chemical kinetic models describing atmospheric chemistry often consider more than fifty individual basic reaction processes.

SO_2 and NO_x are also oxidized within water droplets in clouds. This assertion is based on laboratory measurements which show that SO_2 in solution can be oxidized fairly rapidly by dissolved oxygen, ozone or hydrogen peroxide. For example, using kinetic data from such experiments, calculations indicate that, with ozone, SO_2 and NH_3 levels known to exist in diluted plumes, the oxidation rate is high enough to produce sulphate concentrations, typical of those found in precipitation, within the average lifetime of a cloud droplet (∿ 10 min) (e.g. Cocks and McElroy[17]). The reaction with hydrogen peroxide may be even more rapid (Penkett et al 1979[18]). Pollutants are incorporated into precipitation by 'washout' and 'rainout'.Washout involves scavenging of gaseous or particulate material by precipitation falling through a polluted layer of air and can be easily modelled for SO_2 using known data for solubility (Hales,[12]).

Calculations indicate that, with conditions typical of remote areas washout can account for only a fraction of the sulphate found in precipitation (see Tables 6 and 7) which on average is 70 - 80 μ equivalents per litre (Marsh[13]). This means that much of the sulphate, and indeed the other chemical constituents found in precipitation, must be incorporated in cloud water prior to rainfall.

Table 6 - Showing the result of washout by rain for various SO_2 and NH_3 concentrations in the atmosphere. The initial pH of the rain is assumed to be 5.0 and the rainfall falling from 1000 m (1 mm h^{-1} and 5 mm h^{-1})

Input μg m^{-3}			Output μEquiv/l		
SO_2	NH_3	pH	SO_4^{2-}	NH_4^+	pH
7.5	0	5.0	12	0	4.66
50	0	5.0	44	0	4.27
0	1.0	5.0	0	9	5.85
7.5	1.0	5.0	34	21	4.64
5 mm h^{-1}					
7.5	1.0	5.0	18	11	4.77

Table 7 - Calculated extent of the washout of sulphate aerosol by two different rates of rainfall. The aerosol particles are assumed to be of uniform size.

Sulphate Aerosol $7.5\ \mu g\ m^{-3}$ particle diameter	SO_4^{2-} µEquiv/l in rainfall	
	$1\ mm\ h^{-1}$	$5\ mm\ h^{-1}$
1 µm	7	6
2 µm	22	11

Clearly, cloud processes are of critical importance in the formation of acid rain but better models and, in particular, more measurements are required for different meteorological conditions.

The final process in the formation of acid rain is the precipitation event itself. For rain to occur cloud droplets must grow from a typical average size of ∿ 10 µm to a sufficiently large diameter to fall out by gravity. In temperate latitudes an ice phase is probably necessary because an ice crystal, once formed, can grow rapidly at the expanse of supercooled cloud water droplets at the same temperature because of the lower vapour pressure of the ice.

The composition of rain, including the acidity, varies from place to place and from event to event at any one place. Although most attention has been focussed on specific areas such as S.W. Norway, the acidity of precipitation there is not significantly different from that in, say, the U.K. (see figure 7).

On the basis of a model described schematically in figure 3 , it is possible to calculate annual deposition patterns for sulphur for long range transport using averaged wind fields and simplified assumptions about chemical transformation rates and washout. Figure 8 shows some typical results (Fisher 1978[14]). The regions of high depositions lie mostly near the large industrial emission centres (e.g. Central Europe and the U.K.). The high local deposition rate in S.W. Norway is associated with a higher than average rain (snow) fall in that region. Such calculations agree moderately well with measured deposition rates but they only describe average conditions. In order to attribute observed effects to particular sources it is necessary to develop models to describe transport, transformation and deposition for specific case studies and to verify them by the appropriate atmospheric measurements. The Central Electricity Research Laboratories are currently undertaking a major programme of aircraft flights over the North Sea to make measurements out to distances of several hundred kilometres from source.

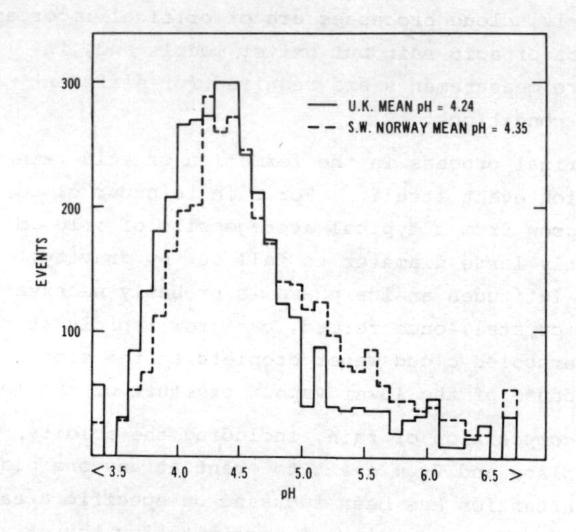

Figure 7 - Comparison of the pH of precipitation at sites in U.K. and S.W. Norway (weighted for equal area).

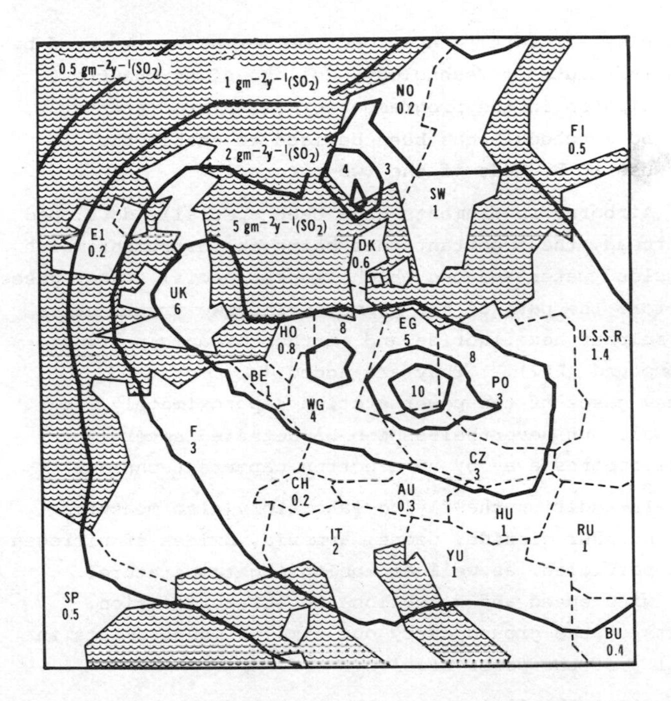

Figure 8 - Calculated total annual deposition of sulphur over Europe ($gm^{-2}\ y^{-1}\ SO_2$) from man-made sources in Europe. Emissions (by country) are given in millions tonnes $SO_2\ y^{-1}$.

To examine the detailed history of the emissions from C.E.G.B. power station chimneys as they move away from the U.K. on the wind a comprehensive set of chemical measurements are being made from a suitably large aircraft. Eggborough power station, near Goole in Yorkshire, was chosen as a typical 2000 MW coal-fired station with 200 m stack sufficiently near to the E. Coast to facilitate tracking over the North Sea. To carry out the measurements, a collaborative programme, co-sponsored by the Electric Power Research Institute, Palo Alto, U.S.A. has been organised with the Research Flight of the Meteorological Office using their large Hercules aircraft. A second smaller aircraft owned by Cranfield Institute of Technology, College of Aeronautics, also participates in the programme for flights of shorter duration and a schedule has been devised to enable the efficient use to be made of the two aircraft.

Airborne instruments have been specially developed at Leatherhead, the important ones being an instrument which collects cloud water and one which tracks special tracer gases and therefore the path of the plume. The two tracer gases used are sulphur hexafluoride and another organic perfluorinated compound (PP2). They are added to the chimney gases of the power station (approximately 1 part per million), but nevertheless can be detected several hundred kilometres away by an electron capture technique.

In addition the flying laboratory also measures traces of sulphur dioxide, ozone, ammonia, oxides of nitrogen and small particles, as well as monitoring temperature, humidity, wind speed and direction and solar radiation. Instruments on the ground carry out similar measurements in the vicinity of the power station.

Since the flying programme started in the Autumn of last year a dozen or so flights over the North Sea out to the Danish and Norwegian Coasts have been completed but the results have not yet been fully analysed.

The C.E.G.B.'s research programme also covers a detailed investigation of the effects of 'acid rain' on ecosystems, again in collaboration with other U.K. and Scandinavian laboratories. The questions being asked relate in particular to the aquatic ecosystems where damage is claimed, the expected time scale of change at present and future levels of deposition, and to distinguish the contribution of rain acidity from other acid inducing processes.

The threshold concentrations of the major phytotoxic gases (SO_2, NO_2 and O_3) for significant economic effects on crop yields are also studied in controlled environment chambers and in greenhouses. Computerised gas control allows the gas concentrations to fluctuate in patterns which reproduce field measurements. It is probable that low concentrations of sulphur dioxide will prove beneficial to crop growth though high levels may be detrimental.

Reverting to Figure 8 it is notable that the extent of the variation in the rate of deposition of sulphur compounds in Europe lies in the wide range 0.5 g/m^2/year to 10 g/m^2/year. The high levels are certainly to do with the burning of fossil-fuel containing sulphur compounds. A recent review of natural and man-made sources of sulphur (Cullis and Hirschler 1980)[15] states that in the Northern Hemisphere (which generates 94% of man-made sulphur) natural sources of sulphur in the atmosphere, geothermal, sea-spray, and the most abundant biogenic emissions, account for 76 million tons a year while man-made emissions are 98 million tons a year. There is clearly a very lively cycle in progress and there is no marked accumulation in the atmosphere though Koide and Goldberg[16] have shown from examination of the sulphur-content of polar ice that the atmospheric concentration has shifted upwards during the last hundred years having previously remained constant over the centuries. The hundred year trend has, of course, accellerated in the last four decades and Cullis and Hirschler[15] estimate that man's global emission of atmospheric sulphur increased from 74.4 million tons in 1965 to 86.1 in 1970 and 104 in 1976.

Coal Cleaning - Chemical measures to minimise the impact of emissions

Over 80% of the coals currently supplied to the Board's power stations experience some degree of cleaning before delivery. The cleaning processes, which are based on differences between the densities and wetting characteristics of the organic matter and other constituents of the raw, run-of-mine coal, are intended to produce a fuel of consistent calorific value and ash content. One can, of course, design cleaning processes specifically to remove pyrites so as to limit sulphur oxide emission. Similarly chloride compounds may be leached from coal thus limiting the release of HCl which is notably an unwelcome emission but also causes particularly severe corrosion on boiler wall tubes when the chloride is present in the coal at a level of 0.3% or more. For this reason washing procedures have been considered for the treatment of high-chlorine coals. Unfortunately the removal of the chloride is a slow process for some of the harder less micro-porous high-chlorine coals. Accellerated chloride removal by washing can be achieved by pretreating the coal with liquid ammonia but the economy of this scheme is not promising.

Sulphur is present in coal in two forms; as pyrites and as organically bound sulphur in the coal substance. The two types are almost equal in an average 1.6 per cent sulphur coal. Removal of pyrites by density separation and froth flotation are well-known processes. They could be applied to remove up to half of the pyrites in a run-of-mine coal - though expensive pregrinding would be necessary to remove that half of the pyrites which is distributed in finer particles within the coal itself.

The weak paramagnetism of the pyrites mineral forms as found in coal has been suggested as a possible means of enhancing gravity separation. It could be applied to dry powdered coal which means that a larger proportion of the pyrites could be separated from the coal than with the run-of-mine coal subject to normal water-based flotation procedures. It is difficult to handle a dry fine coal which has been wetted for density separation. Unfortunately the magnetic susceptibility of the pyrites is really too weak for good practical separation except with very intense magnetic fields and research has been concentrating on pretreatment - e.g. to convert some of the pyrites to FeS, pyrrhotite, which is ferromagnetic or, indeed, partial oxidation to magnetite. Deposition on the pyrites particles of metallic iron by exposure to iron carbonyl has also been reported as a favourable technique.

These radical approaches to pyrites removal can be combined with lithophyllic ash removal to provide a coal so clean that it could be considered for use in an urban setting where ash handling and disposal would be difficult. But chemical treatment to remove the coal-bound sulphur or gasification of the coal and scrubbing the H_2S from the fuel gas would be necessary to remove the whole of the sulphur.

Combustion with additives to retain sulphur

By adding lime or magnesium hydroxide to the combustion gases it is possible to retain a high proportion of sulphur - as calcium or magnesium sulphate in the solid ash. Unfortunately the short residence time available in the boiler means that the added lime must be in the form of very fine (e.g.< 10 μm) particles to ensure that it has time to react to a reasonable extent. An alternative procedure is to burn the fuel in a fluidised bed to which the coarsely crushed lime is added. Experience has shown that limestone or dolomite can be used since there is time to decompose the carbonate as well as for the particles of additive to react in depth with the sulphur oxides. A ratio of Ca/S of 2:1 is said to be sufficient to retain nearly 90% of the sulphur in the coal as calcium sulphate with the ash.

Flue Gas Desulphurisation

Flue gas desulphurisation plants are being installed in existing as well as new power stations particularly in the U.S.A., Japan and West Germany. The technique most often used is to wash the flue gases with a lime slurry. The resultant calcium sulphite,or sulphate if air is blown through it, must then be removed. A regenerable absorbent process based on the $Na_2SO_3/NaHSO_3$ conversion is also possible; it can be operated to produce either sulphur or sulphuric acid as the end-product. A recent study has shown that a number of adaptations to the process as originally developed would be required to cope with U.K. conditions, following, in particular, from the substantially higher chlorine contents of U.K. coals. The chlorine, converted to HCl in the flue gases, interferes with the SO_2 absorption. The removal of the chlorine would, however, create a significant additional need for waste product disposal facilities. Cost studies show that flue gas desulphurisation installed on a new station would add about 15% to the cost of electricity - it would be higher if applied to an existing power station. It is clear that in designing future coal-fired power stations consideration may have to be given to the possibility of cleaning both fuel and flue gas and an optimum procedure developed to obtain the maximum benefit from any extra costs incurred.

Conclusion

The science which went with the evolution of the Tall Chimney to permit the operation of large power stations is the work of physicists, mathematicians and meteorologists. Their studies were based on the desire to ensure that sulphur dioxide emitted from the stack at the rate of 28 tonnes per hour from a 2000 MW station is so dispersed that at ground level it is entirely acceptable and, indeed, does not exceed normal ambient levels. The contribution of the chemist encompases a wide-range of other potentially environmentally unacceptable substances and has extended in recent years to studying environmental impact remote from the source.

Hopefully in this review some indication has emerged of the interesting chemistry involved within the power plant and in the atmosphere. Many topics have been referred to only briefly or not at all. Thus a wide area of chemical research is concerned with the effect of deposited dusts on vegetation and in the soil and there is much more fascinating chemistry and chemical engineering concerned with flue gas washing than it has been possible to mention.

Without doubt the chemist must be prepared to meet challenges which would emerge if coal were to continue to be used as a power station fuel into the next century.

References

1. A.J. Clarke, D.H. Lucas and F.F. Ross, Proc. 2nd Int. Clean Air Conf., Washington, D.C., U.S.A., 1970.

2. C.P. Keeling and R.B. Bacaston, Chapter 4 of National Academy of Sciences Report "Energy and Climate", Washington D.C., 1977

3. J. Billingsley, A.S. Kallend and A.R.W. Marsh, C.E.R.L. Note RD/L/N 121/73, 1973.

4. E. Raask and M.C. Bhaskar, Concrete Research, 1975, 5, 363

5. J. Brown, N.J. Ray and M. Ball, Water Research, 1976, 10, 1115.

6. E. Raask and C. Schilling, Ann. Occup. Hygiene, 1980, 23, 147.

7. R.L. Davison, D.F.S. Natusch, J.R. Wallace and C.A. Evans, Environ. Sci. Technol., 1974, 8, (13) 1107.

8. "A Study of Atmospheric Pollution in Great Britain", Institute of Heating and Ventilating Engineers, 1973.

9. R.D. Smith, J.A. Campbell and K.K. Nielson, Atm. Environ., 1979, 13, (5), 607.

10. W.D. Halstead, Proceedings, Symposium "Coal Conversion and the Environment", Richland, U.S.A., October 1980., Pub. by U.S. Dept. of Energy, Washington.

11. J.M. Allen, A. Levy, P.W. Jones and R.I. Freudenthal, "Polycyclic Organic Materials and the Electric Power Industry", E.P.R.I. Report EA-787-SY, Dec., 1978.

12. J.M. Hales, Atmos. Env., 6, 635, 1972.

13. A.R.W. Marsh, Atmos. Env., 1978, 12, 401.

14. B.E.A. Fisher, Atmos. Env., 1979, 12, 489

15. C.F. Cullis and M.M. Hirschler, Atm. Env., 1980, Vol. 14, pp 1263-1978.

16. M. Koide and E.P. Goldberg, J. Geophys. Res., 1971, 76, 6589-6596

17. A.T. Cocks and W.J. McElroy, C.E.R.L. Note RD/L/N 23/80 1980

18. S.A. Penkett, B.M.R. Jones, E.A. Brice and A.E.J. Eggleton, Atmos. Env., 13, 123, 1980

Electrochemical Energy Storage

By E. J. Cairns
LAWRENCE BERKELEY LABORATORY AND UNIVERSITY OF CALIFORNIA, BERKELEY, CALIFORNIA 94720, U.S.A.

I. Introduction

The energy economies of the world are rather heavily dependent upon petroleum, especially for the generation of electrical energy during peak demand periods, and for transportation. This is shown clearly in Figure 1, (1) which is an energy flow diagram for the U.S. energy economy of 1980, projected from several years earlier. It is evident from Figure 1, and the knowledge that much of the petroleum used by the electric utilities is for generation of power during the peak demand periods, that storage of energy in the utility system can reduce the demand for petroleum, and shift the demand toward other primary energy sources including coal and nuclear energy.

The storage of energy for electric utilities can be done in a number of ways. The most prominent is pumped hydroelectric storage. This requires large, elevated reservoirs, imposing rather severe geographic restrictions on its use. An alternative is to use storage (rechargeable) batteries, which can be installed in almost any location, and have few restrictions. Batteries can also be used to help match the energy supply to the energy demand in wind- or solar-powered electric energy generating systems.

The demand for petroleum used by electric utilities can also be reduced by providing electrical generating systems with the ability to follow the peak load effectively, while operating at

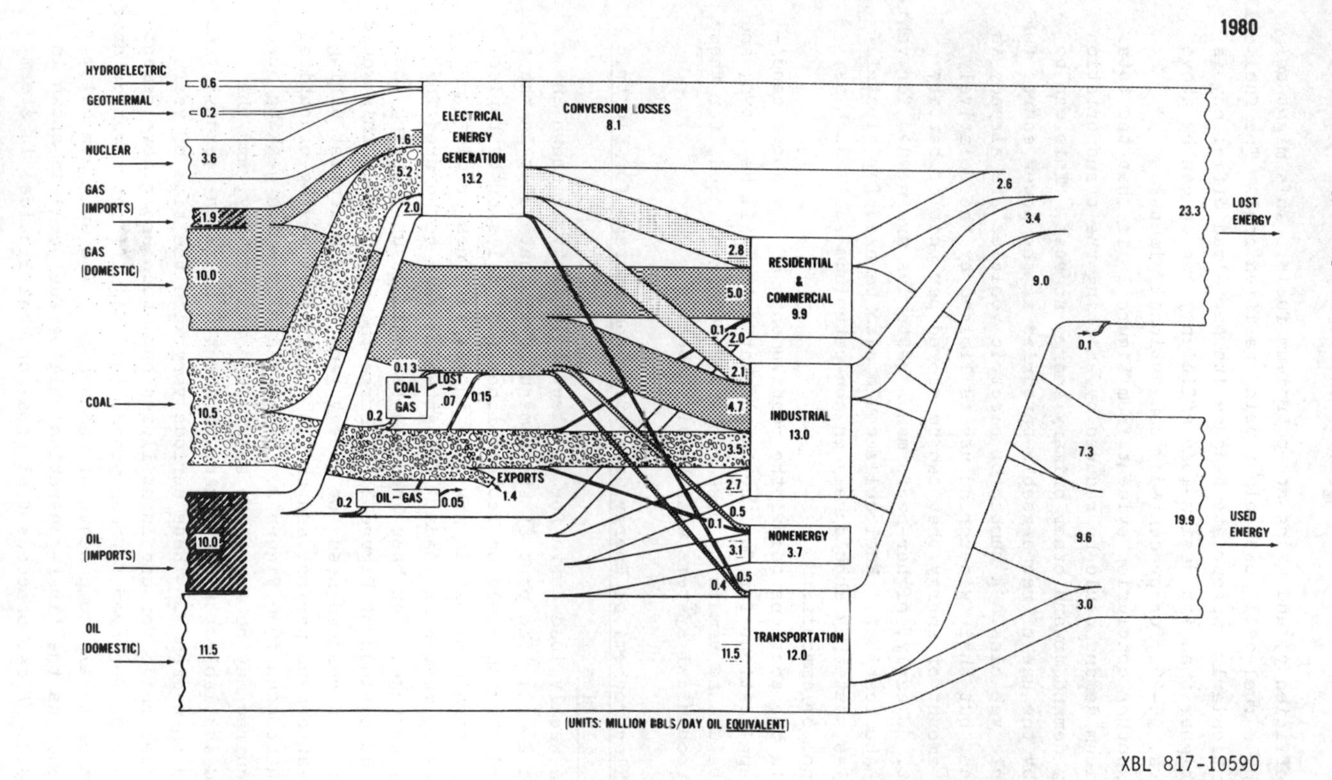

XBL 817-10590

Figure 1. Diagram of the energy economy of the U.S., projected to 1980. The units are expressed in millions of barrels of oil equivalent per day. (1 barrel of oil = 5.8 x 10^6 BTU)

high efficiency, and using non-petroleum fuels. This might be done by a fuel-cell power plant using gasified coal. The fuel cell is particularly appropriate because its part-load efficiency is even higher than its design-load efficiency, which can be above 50% (vs. 35-40% for present-day conventional plants).

Another conclusion evident from Figure 1 is that the total petroleum demand could be reduced by shifting the transportation energy demand toward other primary energy sources. This can be done by the use of rechargeable batteries as the power source for electric vehicles. A number of electric vehicles are already in service, but their widespread use is limited by the relatively small amount of energy that can be stored per unit of battery weight. Clearly, higher-performance batteries are needed for vehicular applications. Fuel cells could also be used for electric vehicles, but they should rely on non-petroleum fuels such as methanol or ammonia.

In the sections below, the requirements of the above candidate applications for fuel cells and batteries will be shown, and the status and remaining research and development needs for these electrochemical systems will be discussed.

II. Electrochemical Energy Generation and Storage for Electric Utilities

A weekly load curve for an electric utility is shown in Figure 2. (2) The part of the demand curve that falls in the zone labeled peaking could be met by either fuel cells or batteries, saving petroleum that is now used in gas turbines or diesel engines to meet peak loads. If batteries were to be used for energy storage, then base load plants could be used during the low-load periods of Figure 2 to charge the batteries. This would result in a reduced need for energy generation equipment during the peak demand period, and should result in a saving of capital. As can be seen from Figure 2, the time during which peaking power is required may be from 3 to about 10 hours per day, and the period available for recharging is about 5 hours on weekdays, and longer on weekends. These periods vary with the time of year.

Fuel cells for use in utilities have been under development in the U.S. for over fifteen years. (3) Significant advances have been made in all components of the system. The approach being developed as the first-generation utility fuel cell is shown in Figure 3. A carbonaceous fuel (coal or coal liquids) is steam-reformed to produce a hydrogen-rich fuel stream that is fed to a

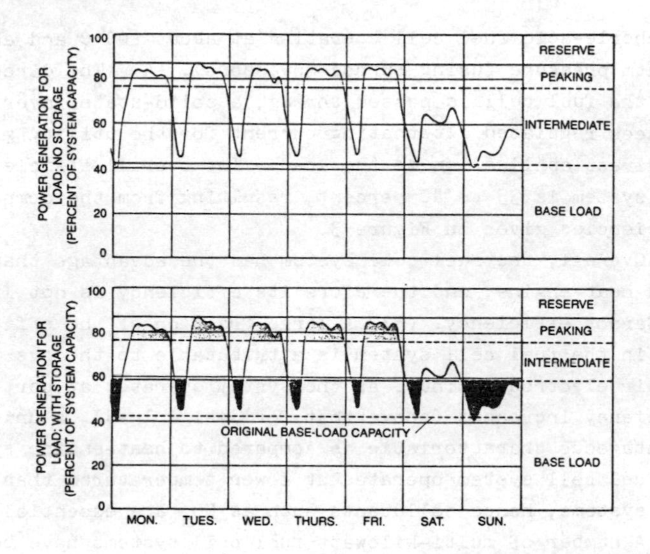

XBL 818-10953

Figure 2. Example of a weekly load curve for an electric utility, showing energy available for storage during off-peak periods (dark shading), and energy that could be supplied by storage batteries.

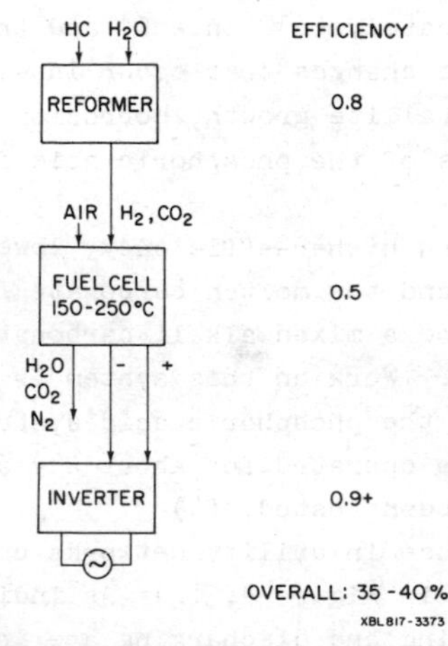

Figure 3. Schematic diagram of a fuel-cell system.

phosphoric-acid fuel cell operating at about 200°C and about 3.5 atm pressure (using turbocompressors). (4) The direct current from the fuel cell is passed through a solid-state inverter, which produces regulated alternating current for the utility grid, or for direct application to the load. The overall efficiency of this system is 35 to 40 percent, resulting from the component efficiencies given in Figure 3.

Overall, the fuel cell system has the advantage that it is not a heat engine, and therefore its efficiency is not limited by the Carnot efficiency. (5) A large fraction of the efficiency loss in the fuel cell system is attributable to the overvoltage of the air electrode. Thus, as the system operates at part load, its efficiency increases (above that for design load), a unique and advantageous characteristic as compared to heat-engine systems. The fuel cell system operates at lower temperatures than combustion systems; hence pollutants such as NO_x are essentially absent.

A number of multi-kilowatt fuel cell systems have been tested, and lifetimes with maintenance and cell stack exchange have exceeded 10,000 hr. Scale-up of cells and components has taken place, and stacks of 20-24 cells sized for use in 4.8 MW systems have demonstrated lifetimes up to 14,000 hr. A short test of a 1 MW system has taken place, and two 4.8 MW systems are now under construction for testing in the U.S. and Japan. (4)

Although the advantages of fuel cell systems are very attractive, a number of problems remain to be solved before they can be commercially successful. Acceptable cost and lifetime are the key issues. The cost is related to the precious metals (platinoid elements as electrocatalysts) content, and the life is related in a complex way to the changes that occur in electrocatalysts with continued use (crystallite growth, corrosion of the graphite substrate). The status of the phosphoric acid fuel cell system is shown in Table 1.

Advanced work on higher-efficiency, lower-cost fuel cell systems centers around the molten carbonate fuel cell, which uses nickel electrodes and a mixed alkali carbonate electrolyte, and operates near 700°C. Work on this system is at a much earlier stage than that for the phosphoric acid system. Single molten carbonate cells have operated for about 2 years and some stacks of 900 cm^2 cells have been tested. (4)

Batteries for use in utility networks could be arranged as shown schematically in Figure 4. (2) As indicated above, the times available for charging and discharging are in the range of 3-8

Table 1

$H_2(Pt)/H_3PO_4/(Pt)$Air Fuel Cell System

Advantages

Good efficiency for small systems (35-40%)
Efficiency increases for lower loads
Low pollution from reformer and cell

Status

Multi-kW systems tested - over 10,000 hr with maintenance and stack exchange
Short tests of 1-MW system completed
4.8 MW systems under construction

Problems

Performance decay-catalyst sintering
High cost
Short life
Efficiency too low for base load systems

Recent Work

Pressurized operation for higher performance
Continued catalyst development
Lower-cost cell parts
Large cells (0.35 M^2) and systems (1 MW)

hours. In order to be competitive with pumped hydroelectric energy storage, which has an efficiency of 65-70%, the battery system should be at least 70% efficient. The battery system would require a building, and the tolerable cost of the building is such that the battery should store at least 80 kWh of energy per square meter of floor area (with a height of less than 6.1 meters: this corresponds to 30 Wh/ℓ). For a significant-sized energy-storage substation, the battery should store 100-200 MWh of energy. If the battery were to cost $30/kWh and last for 2000 deep cycles, (perhaps 200 cycles/yr for 10 years) then this would amount to 1.5¢/kWh of energy stored. In some areas significantly higher costs can be tolerated. These requirements are summarized in Table 2.

If the requirements of Table 2 are met, then batteries would have an economic advantage over almost any other means of energy storage for discharge times of up to about 10 hours per day. At higher costs (up to about $100/kWh), batteries would still have an advantage for shorter discharge times. A detailed discussion of the various economic and other trade-off considerations for

Table 2

Requirements for Off-Peak Energy Storage Batteries

Discharge time	3-8 hours
Charge time	5-7 hours
Overall efficiency	>70%
Energy/floor area (6.1 m max. height)	80 kWh/m^2
Typical size	100-200 MWh
Cycle life	2000
Lifetime	10 years
Cost	$30/kWh

batteries versus other energy storage and generation technologies is presented in Reference 6.

The electric utilities in the U.S. are seriously evaluating the viability of energy storage in batteries. A Battery Energy Storage Test facility (BEST facility) has been constructed in Somerset County, New Jersey for the testing of battery modules, up to 5 MWh each, in regular utility service. A photograph of a model of this facility is shown in Figure 5. (7) The first battery to be used in evaluating the operation of the facility is a $Pb/H_2SO_4/PbO_2$ battery; the second one will probably be $Zn/ZnC\ell_2/C\ell_2 \cdot 8H_2O$. These tests will be taking place in 1981 and following years.

In order to gain perspective with regard to the present status of and future prospects for batteries in electric utility systems, it is useful to review some of the individual candidate batteries for this application. As a baseline for comparison, the $Pb/H_2SO_4/PbO_2$ system is most appropriate, since it is the only commercially-available battery that approaches most of the performance, life, and cost requirements.

The status of the Pb/PbO_2 cell is shown in Table 3. For utility applications, requiring long cycle life, it is possible to obtain 1500-2000 cycles, but at a low specific energy of about 20 Wh/kg. The cost of these long-lived batteries is greater than that shown in Table 3 (for vehicle batteries) and may be about $125-150/kWh. Recently, maintenance-free cells have been developed in automotive sizes, and it is expected that this feature will be employed in larger cells. If large Pb/PbO_2 battery systems are to be used, it probably will be a great benefit to develop sealed cells, with internal recombination of the gases. This has already been done in very small sizes (several Ah).

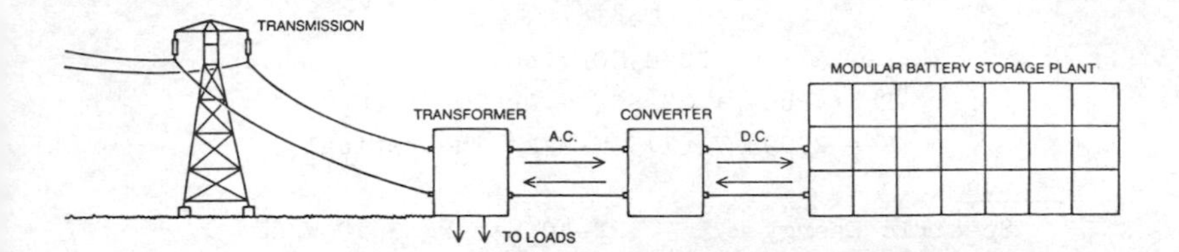

XBL 818-10952

Figure 4. Schematic diagram of how batteries might be used for energy storage in a utility network.

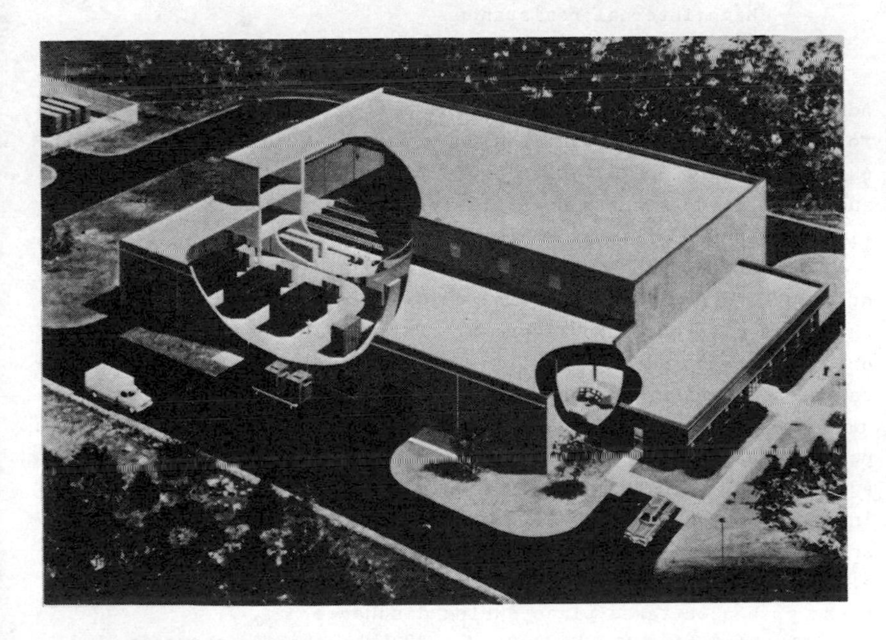

Figure 5. Battery Energy Storage Test Facility, Sommerset County, New Jersey.

Table 3

$Pb/H_2SO_4/PbO_2$

$Pb + PbO_2 + 2H_2SO_4 \rightarrow 2PbSO_4 + 2H_2O$

E = 2.095 V; 175 W·h/kg Theoretical

Status

Specific Energy	22-40 W·h/kg @ 10 W/kg
Specific Power	50-100 W/kg @ 10 W·h/kg
Cycle Life	300+ @ 10 W/kg, 60% DOD
Cost	$50/kW·h

Recent Work

- Replace Sb with Ca in positive current collector
- Maintenance-free cells
- Use $4PbO \cdot PbSO_4$ instead of $PbO + Pb_3O_4$ in positives
- New, low-resistance current collectors

Problems

- Sealing of cells
- Positive current collector corrosion
- Cohesion and adhesion of PbO_2
- High internal resistance
- Heavy

Advances have also been made in the design of current collectors for minimum cell resistance, using computer-aided design techniques. The cost and lifetime projections for Pb/PbO_2 cells are such that it does not appear likely that this battery will be widely used for off-peak energy storage.

A system which may prove to be economically acceptable for off-peak energy storage is zinc/chlorine. A schematic diagram of this system is shown in Figure 6. This is a flow system, with the chlorine stored as $C\ell_2 \cdot 8H_2O$, an ice-like solid, in a separate compartment. During the charging process, zinc is deposited on the dense graphite negative electrodes, while chlorine is evolved from porous graphite positive electrodes. The chlorine-saturated aqueous $ZnC\ell_2$ electrolyte is circulated through a chiller, bringing its temperature below 9° C, where the $C\ell_2 \cdot 8H_2O$ forms. The solid $C\ell_2 \cdot 8H_2O$ is filtered out in the storage area, and the $ZnC\ell_2$ electrolyte is recirculated to the cell stacks. The reverse of these processes takes place during discharge.

The status of the $Zn/ZnC\ell_2/C\ell_2 \cdot 8H_2O$ system is shown in Table 4. (8) With a theoretical specific energy of 405 Wh/kg, it is possible that a practical specific energy of 80-90 Wh/kg might be achieved (66 Wh/kg has already been demonstrated). In a small

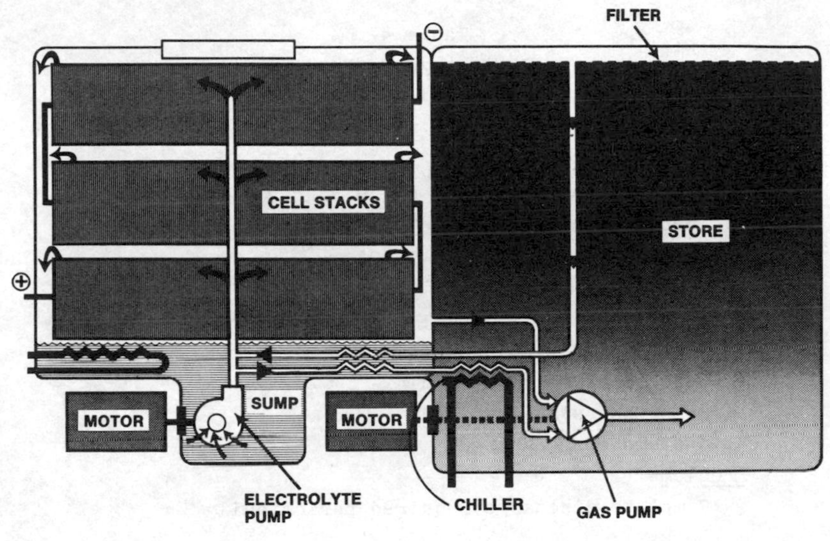

Figure 6. Diagram of the zinc/chlorine system. (Courtesy of Energy Development Associates)

system, a cycle life of 1400 cycles, with electrolyte maintenance (for purity) has been achieved. Systems as large as 50 kWh have been built, and will probably serve as modules for the BEST facility battery to be tested in 1981 or 1982.

Some of the remaining problems of the $Zn/C\ell_2 \cdot 8H_2O$ system to be addressed include the necessity of periodically discharging all of the zinc in order to avoid severe dendrite formation, shorting the cells. Additives are used in the electrolyte to help control the morphology of the zinc deposit, and minimizing the frequency of complete discharge. It has been found that low concentrations of iron (in the ppm range) and some other metals in the electrolyte result in excessive hydrogen evolution from the zinc electrode, so maintenance of a high-purity electrolyte (with regard to low H_2 overvoltage metals) is important. Recombination of the H_2 with $C\ell_2$ is promoted by an ultraviolet light source in the gas space above the cells. Because this system does not use separators, the dissolved $C\ell_2$ remaining in the electrolyte after it passes through the porous graphite positive electrode may combine directly with the zinc on the negative electrode, resulting in an efficiency loss. Consequently, the $Zn/C\ell_2 \cdot 8H_2O$ system operates at

Table 4

$Zn/ZnCl_2/Cl_2 \cdot 8H_2O$

$Zn + Cl_2 \cdot 8H_2O \rightarrow ZnCl_2 + 8H_2O$

E = 2.12 V; 405 W·h/kg Theoretical

Status	
Specific Energy	66+ W·h/kg @ 3-4 W/kg
Specific Power	70 W/kg for seconds
Cycle Life	1400*
Cost	>$100/kW·h

Recent Work

- Additives for Zn deposition
- Recombination of H_2 and Cl_2
- 35-50 kWh systems
- Systems components

Problems

- Complete discharge required periodically
- Bulky
- Complex
- Low specific power
- Very sensitive to impurities
- Low efficiency
- Gaskets

*1 kWh system only, with electrolyte maintenance

about 60-70% efficiency, which could be a problem. Work continues on the improvement of the system as 50 kWh units are prepared for test. Cost projections fall below $100/kWh, and make this system potentially attractive for stationary energy storage.

A high-temperature cell which has been under development for about fifteen years, and is a candidate for utility energy storage, is the sodium/sulfur cell, which uses a $Na_2O \cdot 9Al_2O_3$ (beta alumina) ceramic electrolyte, and operates at 350°C. The ceramic electrolyte is used in the form of a closed-end tube, and has one reactant inside, the other outside, as shown in Figure 7. (9) The sulfur is held in the pores of a graphite felt current collector, and the cell must be sealed in order to avoid reaction with air and moisture. A number of batteries (about 10 kWh) have been built for demonstration purposes, but none of these have shown a significant life.

Some discharge and charge curves for a parallel-connected Na/S battery of about 25 cells (∿120 Ah each) are shown in Figure 8. (10) The lifetimes of individual cells in the

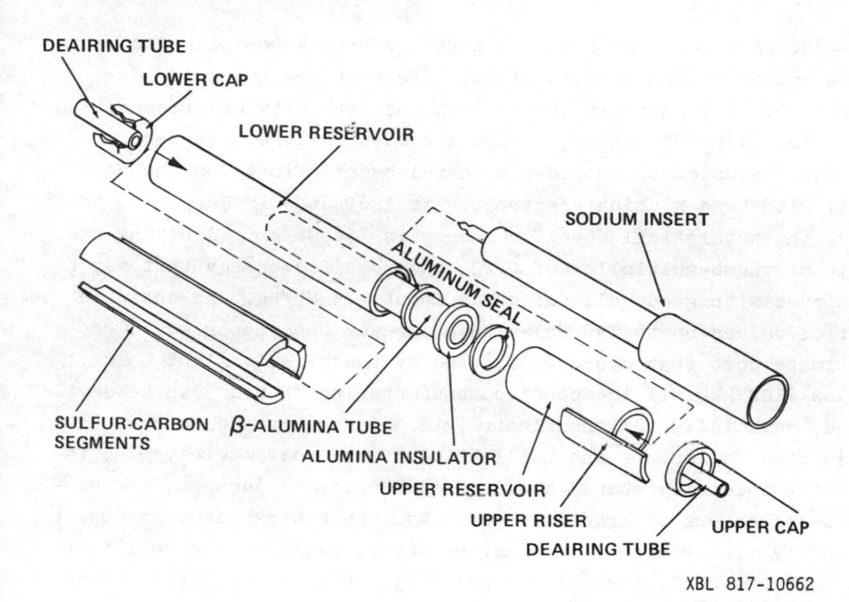

Figure 7. Exploded view of a Na/S cell.

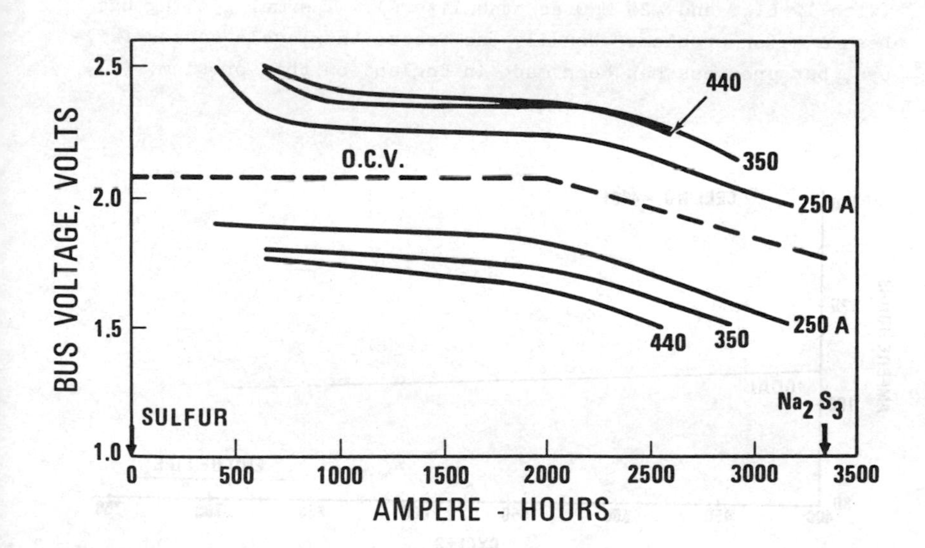

Figure 8. Charge and discharge curves for a 25-cell parallel-connected Na/S battery. (10)

100-200 Ah size range has been highly variable (between 200 and 1500 cycles). The results of one life test are given in Figure 9, (10) where it can be seen that capacity maintenance was good for over 700 cycles, before a sudden failure occurred.

The status of the world-wide development efforts on the Na/S cell with beta alumina electrolyte is indicated in Table 5. Notice that the theoretical specific energy is 758 Wh/kg. Applying the rule of thumb multiplier of 0.23, the specific energy that may be achieved with good cell design is about 175 Wh/kg. As shown in the table, values up to 140 Wh/kg have already been reported. The ultimate cost that could be reached by these cells might be well below $100/kWh, if inexpensive manufacturing methods can be developed, especially for the electrolyte. Innovative approaches have been used to improve the utilization of the sulfur electrode, including specially shaped and layered current collectors, and current collectors of graded resistivity. Much work has been done to identify electronically conductive materials which are resistant to sulfur attack, including doped TiO_2. New electrolytes having higher conductivity at lower temperatures have been sought, including Nasicon ($Na_{1+x}Zr_2Si_xP_{3-x}O_{12}$), but beta" alumina remains best (with 1% Li_2O and ∿2% MgO as stabilizers). Thermal cycling has been a major problem. Usually one freeze-thaw cycle causes failure, but progress has been made in England on this problem. (11)

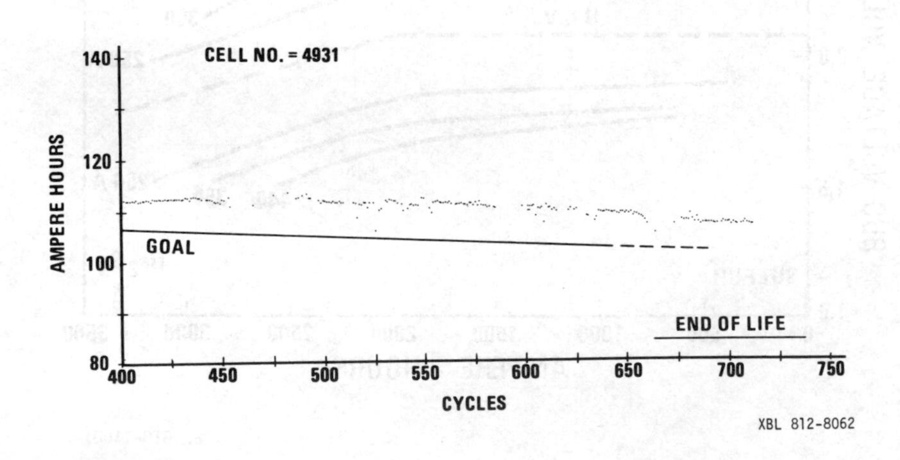

Figure 9. Capacity vs. cycle number for a full-sized Na/S cell. (10)

Table 5

Na/Na+ Solid/S

$2Na + 3S \rightarrow Na_2S_3$

$\bar{E}$ = 2.0 V; 758 Wh/kg Theoretical

Status	
Specific Energy	85-140 Wh/kg @ 30 W/kg
Specific Power	60-130 W/kg peak
Cycle Life	200-1500
Lifetime	3000-15,000 h
Cost	>$100/kWh
Recent Work	
Batteries, ~10 kWh	
C_6N_4 additive to S	
Ceramic (TiO_2) electronic conductors	
Shaped current collectors	
Tailored resistance current collectors	
Sulfur-core cells	
$Na_{1+x}Zr_2Si_xP_{3-x}O_{12}$	
Thermocompression bonded seals	
Problems	
Corrosion-resistant material for contact with S	
Low cost seals	
Low cost electrolyte	
Specific power is low	
Thermal cycling	

Another high-temperature cell under consideration for energy storage in electric utilities is the LiAℓ/FeS cell, which uses a molten-salt electrolyte of LiCℓ-KCℓ, and operates at 450° C. A cutaway view of this type of cell is shown in Figure 10. (12) The electrodes are prepared from mixtures of powdered salt electrolyte and powdered reactant (LiAℓ, FeS) pressed into a plaque, and assembled with current collectors, particle retainers, and DN felt separators as shown in Figure 10. The cell is sealed to prevent reaction with air and moisture.

Typical discharge curves for a LiAℓ/LiCℓ-KCℓ/FeS cell are shown in Figure 11. (13) This cell had a capacity of 80-90 Ah, but cells currently under test have capacities of about 350 Ah because they contain a number of electrodes internally connected in parallel. The cycle life of such cells is about 350 cycles; the specific energy of the most recent cells of this general type has been about 100 Wh/kg.

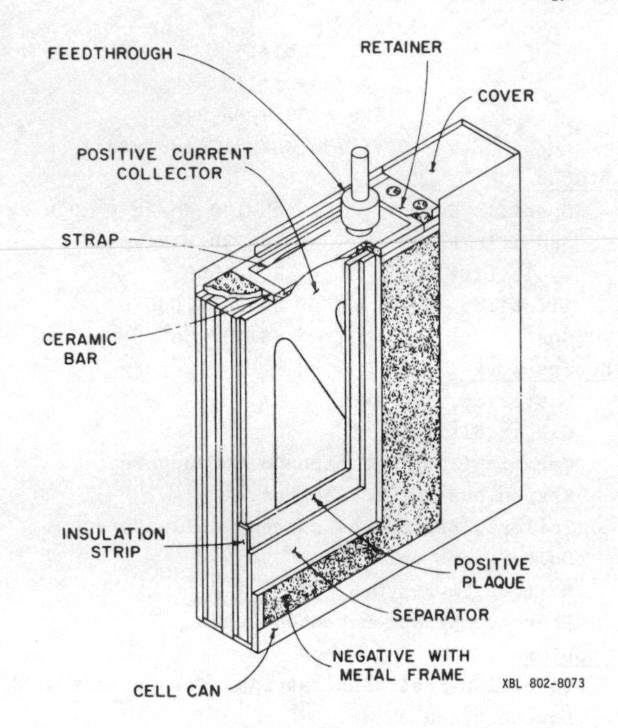

Figure 10. Cutaway drawing of a LiAℓ/FeS cell.

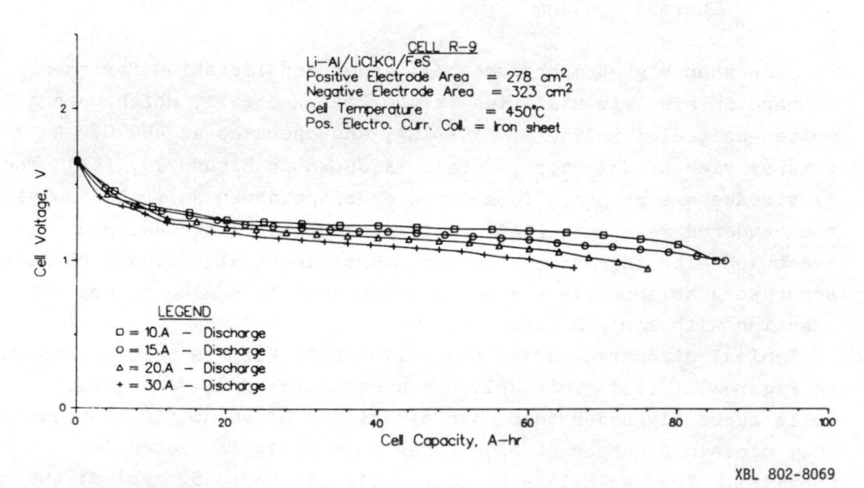

Figure 11. Discharge curves for a LiAℓ/FeS cell operated at various constant currents, as labeled.

Recent work on LiAℓ/FeS cells has included improvements in the wetting of the BN separator by the electrolyte, improved current collectors for lower internal resistance and higher specific power, and cost reduction measures for the BN separator. Electrical shorting of the cells, due to extrusion of the positive electrode active material and protrusions of the negative electrode active material, remains the main failure mode. Recent progress has been made on tolerance to thermal cycling. Up to 30 freeze-thaw cycles can be experienced by these cells without capacity loss. The status of the LiAℓ/FeS cell is summarized in Table 6.

Table 6

LiAℓ/LiCℓ-KCℓ/FeS

$$2LiA\ell + FeS \rightarrow Li_2S + Fe + 2A\ell$$

E = 1.33 V; 458 Wh/kg Theoretical

T = 450°C

Status	
Specific Energy	60-100 Wh/kg @ 30 W/kg
Specific Power	60-100 W/kg, peak
Cycle Life	300+ @ 100% DOD
Lifetime	5000+ h
Cost	>$100/kWh

Recent Work

- Multielectrode cells
- LiX-rich electrolyte
- BN felt separators
- Wetting agent for separators
- Powder separators-MgO
- Batteries of 320 Ah cells
- Improved current collectors

Problems

- Low specific energy
- Low voltage per cell
- Cell shorting major failure mode
- Electrode swelling and extrusion
- Agglomeration of Li-Aℓ with cycling
- Capacity loss
- High separator cost
- Leak-free feedthroughs
- Thermal control

Some conceptual design work has been performed on utility energy storage modules. Figure 12 shows a drawing of a truckable utility module, which would have an energy storage capability of about 5.6 MWh, and would contain its own thermal control system and electronics. (14)

It is clear from the discussion above that there are several battery systems under development that may provide for storage of a significant amount of energy for utility systems. Evaluations in the BEST facility during the next few years should provide guidance to the selection of the most appropriate batteries for this application.

III. Energy Storage for Solar- and Wind-Powered Systems

During the last several years, the idea of solar- and wind-powered electrical systems has gained in popularity because of the desire to decrease dependence on petroleum and other polluting energy sources. Unfortunately, solar and wind energy are not necessarily available when energy is needed. This introduces the need for an energy storage system which may be required to provide

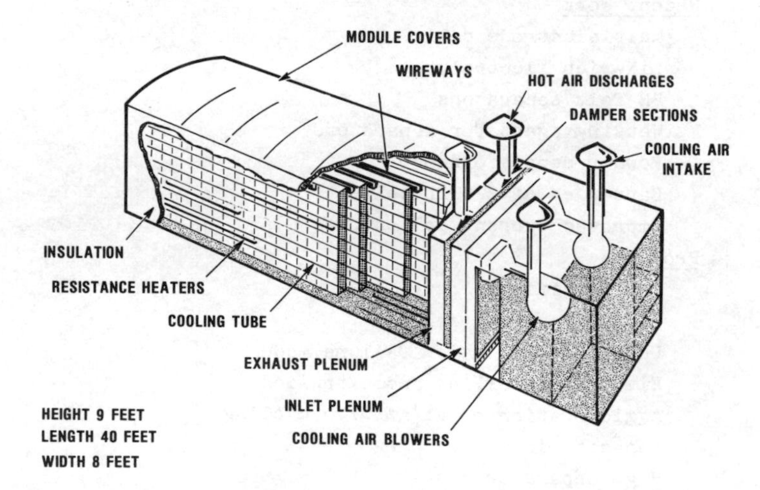

Figure 12. Schematic cutaway drawing of a LiAℓ/FeS truckable battery module for possible use in electric utility networks. (14)

energy for periods of 12 hours to a few days. Because of this extended energy delivery period, flow batteries with storage of reactants in tanks are the most popular concept for this application. In this way, the energy storage part of the system can be at least partially decoupled from the energy conversion part of the system.

The $Zn/ZnC\ell_2/C\ell_2 \cdot 8H_2O$ system has been proposed for the solar and wind energy storage application. The status of this system was discussed above. A related system is $Zn/ZnBr_2/Br_2$, in which the bromine is stored as a chemical complex in a separate tank. A few, relatively small multi-kWh systems have been built and tested. These systems have efficiencies in the 60-70% range. The cells make use of an ion-exchange membrane to prevent direct reaction between Br_2 and Zn. (15) Another system making use of the Zn electrode is the $Zn/Fe(CN)_6^{-3}$ cell, which also contains an ion exchange membrane, and uses external storage of the ferricyanide solution. Finally, there is under investigation a system in which both reactants (a Cr^{+2} solution, and a Fe^{+3} solution) are stored in tanks. All of these systems are less well developed than those discussed in the previous section of this paper, and it is probably premature for a detailed report.

IV. Energy Storage for Electric Vehicles

The specific power and specific energy requirements for electric vehicle applications can be assessed by applying the equations of motion for vehicles to typical driving profiles. This has been done as discussed in References 16 and 17. The results of those calculations can be summarized conveniently in the form of Table 7. A useful value for urban vehicle battery energy calculations is 0.15 kWh/T-km. The range that can be expected from an electric vehicle operating on an urban driving profile is:

$$R(km) = \frac{SpE(Wh/kg)}{0.150(Wh/kg\text{-}km)} \times \frac{Mb(kg)}{Mv(kg)} \qquad (1)$$

where SpE is the specific energy of the battery

Mb is the mass of the battery

Mv is the test mass of the vehicle.

It can be seen from Equation 1 that for a range of 150 km, and Mb/Mv = 0.3 (the maximum for good automotive design), the specific energy must be at least 75 Wh/kg. Because of the fact that there are no batteries available with a specific energy of 75 Wh/kg, a reasonable cycle life (>300 cycles) and an acceptable cost

Table 7
Energy and Power Requirements
for Urban Electric Vehicles

<u>Energy Consumption</u>	
At axle:	0.10-0.12 kW·h/T·km
From battery:	0.14-0.17 kW·h/T·km
From plug:	0.18-0.23 kW·h/T·km
<u>Peak Power Required</u> (0-50 km/h, ⩽10 s)	
At axle:	25 kW/T (Test wt.)
From battery:	35 kW/T (Test wt.)
<u>Average Power Required at Axle</u>	
Fed. Register	∿5 kW/T (Test wt.)
50 km/h cruise	∿3 kW/T (Test wt.)
Peak for 0-50 km/h, ⩽10 s	∿25 kW/T (Test wt.)

<$100/kWh, there has been a significant effort to develop advanced batteries for electric vehicles.

The battery that is closest to meeting the performance, cycle life and cost goals for electric automobiles is zinc/nickel oxide. The electrolyte is aqueous potassium hydroxide (∿33 w/o). This cell has a theoretical specific energy of 326 Wh/kg, and should therefore be capable of providing 75 Wh/kg in a practical configuration. Values of 60-80 Wh/kg have already been demonstrated in batteries up to full electric vehicle size, such as the one shown in Figure 13.

The zinc/nickel oxide batteries that have been tested in electric vehicles have demonstrated the expected high performance and improved range (over Pb/PbO_2). Unfortunately Zn/NiOOH cells have not delivered acceptable cycle lives, typical values being only 100 to 200 deep cycles. The main cause of short cycle life has been failure of the zinc electrode and/or the separator. Significant efforts have been devoted to improvements in the zinc electrode, but only gradual gains have been realized. Efforts continue on the zinc electrode, and the separator, as well as on cost reduction, especially for the NiOOH electrode. Because of the fact that Zn/NiOOH cells can be operated in the sealed state, they show promise of being developed as totally maintenance-free devices, with internal provision for gas recombination. The status of this cell is summarized in Table 8.

Table 8

Zn/KOH/NiOOH

$$Zn + 2NiOOH + H_2O \rightarrow ZnO + 2Ni(OH)_2$$

E = 1.735 V; 326 W·h/kg Theoretical

Status	
Specific Energy	60-75 W·h/kg @ 30 W/kg
Specific Power	200-300 W/kg @ 35 W·h/kg
Cycle Life	100-200 @ 25-50 W/kg, 60% DOD
Cost	>$100/kW·h
Recent Work	
Inorganic separators (e.g., ZrO_2, $Ni(OH)_2$, $Ce(OH)_3$, others)	
Sealed cells	
Nonsintered electrodes	
Problems	
Sealing of cells - O_2 evolution and recombination	
Zn redistribution	
Separators	

A high-temperature cell that could be developed as a power source for electric vehicles is the Li_4Si/FeS_2 cell, which uses LiCℓ-KCℓ molten salt electrolyte, and operates at 450° C. Because it has a theoretical specific energy of 944 Wh/kg, there is the expectation that cells capable of 200 Wh/kg can be developed. A schematic cross section of a laboratory Li_4Si/FeS_2 cell is shown in Figure 14. It can be seen that the structure of this cell is similar to that of the LiAℓ/FeS cell, discussed above. Cells like that of Figure 14, in 70-80 Ah sizes, have yielded the performance data shown in Figure 15. (18) Note that a specific energy of over 180 Wh/kg was realized.

The status of the Li_4Si/LiCℓ-KCℓ/FeS_2 cell is given in Table 9. Cell lives of about 2 years have been achieved, and specific energies of 120 Wh/kg have been maintained at a specific power of 30 W/kg (typical average specific power for urban driving). These cells must be scaled up to larger sizes and incorporated into thermally self-sustaining batteries before they can be tested in vehicles. Some of the problems that remain to be solved include: inexpensive, corrosion-resistant materials for the current collector in the FeS_2 electrode (Mo and graphite are used now), lower-cost separators (BN felt is used now), and low-cost leak-free feedthroughs.

Figure 13. Photo of a zinc/nickel oxide battery for an electric automobile.

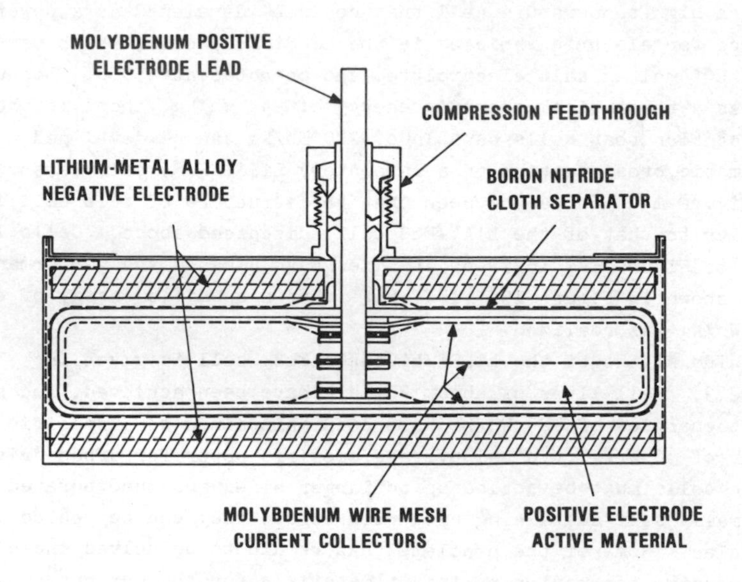

Figure 14. Schematic cross-section of a Li_4Si/FeS_2 cell.

Table 9

$Li_4Si/LiC\ell-KC\ell/FeS_2$

$Li_4Si + FeS_2 \rightarrow 2Li_2S + Fe + Si$

E = 1.8, 1.3 V; 944 Wh/kg Theoretical

Status	
Specific Energy	120 Wh/kg @ 30 W/kg
	180 Wh/kg @ 7.5 W/kg
Specific Power	100 W/kg, peak
Cycle Life	700 @ 100% DOD
Lifetime	~15,000 h
Cost	>$100/kWh
Recent Work	
Bipolar cells	
Li-Si electrodes	
BN felt separators	
70 Ah cells	
Problems	
Materials for FeS_2 current collector	
Leak-free feedthroughs	
High internal resistance	
Low-cost separators needed	
Thermal control	

In order to bring some perspective to the assessment of candidate batteries for electric vehicles, some design calculations were performed, based on a 1000 kg urban vehicle, of the type shown in Figure 16. Allowance was made for the added battery mass when making energy consumption comparisons to the gasoline reference vehicle. The vehicles were designed for a 160 km range, except for the Pb/PbO_2 version, which had a design range of 75 km. For each vehicle, the primary energy consumption was calculated, referring to resource in the ground, using efficiency values given in Reference 17. As can be seen from the results in Table 10, the higher specific energy batteries offer the opportunity to conserve energy, as well as to shift energy demand away from petroleum. There is thus a strong incentive to develop high specific energy batteries for widespread use in urban automobiles. In addition to high specific energy, it is necessary that electric vehicle batteries have acceptable durability and cost. Recent goals for performance, durability, and cost for electric vehicle batteries are given in Table 11. Finally, Figure 17 presents a comparison of the specific power vs. specific energy plots for a number of

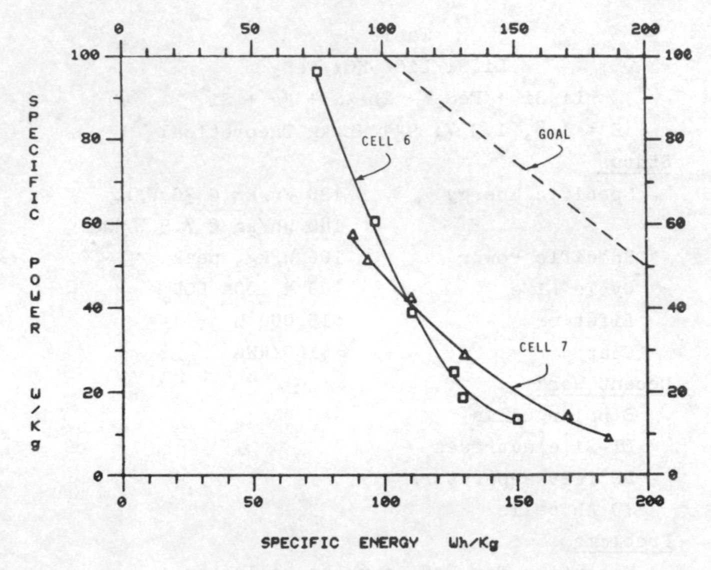

Figure 15. Specific power vs. specific energy for two Li_4Si/FeS_2 cells. (18)

Figure 16. Photo of General Motors Electrovette, an urban electric automobile. (Courtesy of General Motors)

Table 10
Effectiveness and Primary Energy
Consumption Comparison
Urban Autos, 135 kg Payload

	GVM	$\frac{kW \cdot h^{\dagger}}{km}$	Primary Energy Consumption		
			Petroleum	Coal	Nuclear
	kg		kW·h/km	kW·h/km	kW·h/km
Gasoline	1050	0.86	0.92	1.34	3.04
Pb/PbO_2*	1620	0.38	1.32	1.21	1.32
Zn/NiOOH**	1400	0.32	1.14	1.05	1.14
$Li\text{-}Si/FeS_2$**	1100	0.25	0.90	0.82	0.89

†Energy input to vehicle
*75 km range
**160 km range

Table 11
Battery Goals

BATTERY		PERFORMANCE*			DURABILITY			COST†		
Type	Mass (kg)	Specific Energy (W·h/kg)	Energy Stored (kW·h)	Urban Range (km)	Cycles 100% DOD	Years	km	$/kW.h	$/Battery	$/km
Lead-Acid (Advanced)	300	30	9	60	400	3	24,000	50	450	0.019
Zinc/Nickel Oxide (3-5 Years)	250	75	18.75	125	300	3	37,500	70	1300	0.035
Lithium/Iron Sulfide (10+ Years)	150	200	30	200	1000	3-5	200,000	40	1200	0.006

*Basis 1000 kg vehicle, 0.15 kW·h/T·km
†Battery amortization cost only;
Electricity Cost 0.25 kW·h/km x $0.04/kW·h = $0.01/km

battery systems. The high-temperature batteries offer the highest performance, but they still fall short of the spark ignition engine plus fuel tank: an incentive for future work!

V. Conclusions

The discussion above and the data presented allow the following conclusions to be drawn.

1. Electrochemical energy conversion systems offer opportunities to decrease our dependence upon petroleum through more efficient energy conversion, and through energy storage in electric utility networks.

2. Batteries of higher specific energy (>70 Wh/kg) may provide useful electric automobiles, decreasing our dependence on petroleum for transportation.

3. The usefulness of solar- and wind-powered systems is increased by the storage of energy in batteries.

4. Battery lifetime and cost require improvement to meet the needs of the above applications.

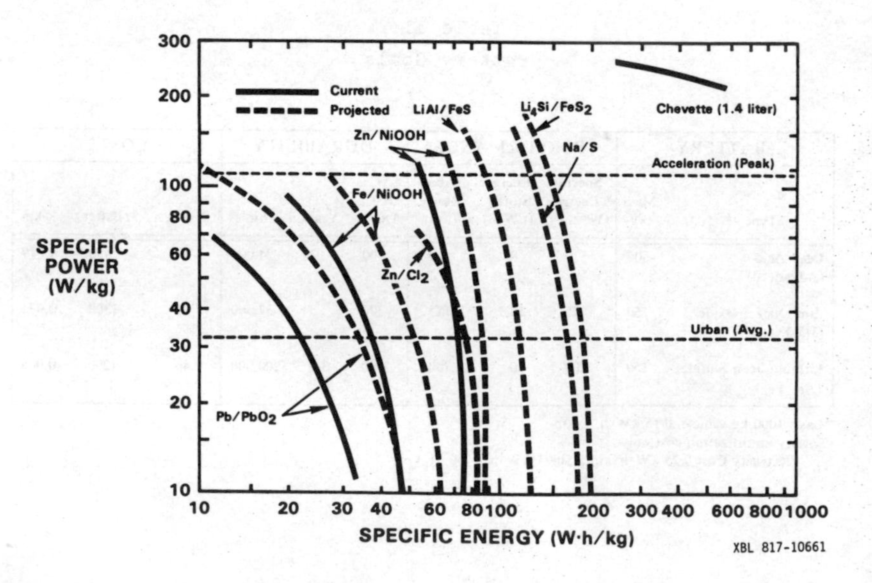

Figure 17. Specific power vs. specific energy plot for various batteries.

REFERENCES

1. A.L. Austin, B. Rubin, and G.C. Werth, Lawrence Livermore Laboratory Report UCRL 51221, May 30, 1972.
2. F.R. Kalhammer, Scientific American, 241, 56, December 1979.
3. S. Orlofsky, in Preprints of Papers Presented at the ACS Division of Fuel Chemistry Biennial Fuel Cell Symposium, Chicago, IL, 1967, p. 309.
4. United Technologies Corporation, Advanced Technology Fuel Cell Program, EPRI Report - EPRI EM-1730-SY, March 1981.
5. H.A. Liebhafsky and E.J. Cairns, Fuel Cells and Fuel Batteries, John Wiley & Sons, Inc., New York, 1968, Chapter 3.
6. M.L. Kyle, E.J. Cairns, and D.S. Webster, Argonne National Laboratory Report ANL-7958, March 1973.
7. E.J. Cairns and R.R. Witherspoon, Batteries, Primary--Fuel Cells, in Kirk-Othmer Encyclopedia of Chemical Technology, John Wiley & Sons, Inc., New York, 1978.
8. C.H. Chi, P. Carr and P.C. Symons, in Proceedings of the 14th IECEC, Volume 1, American Chemical Society, Washington, D.C., 1979, p. 692.
9. General Electric Review of the Advanced Battery Development Program for Electric Utility Application, May 1979.
10. Ford Aerospace & Communications Corporation, Annual DOE Review of the Sodium-Sulfur Battery Program, April 1980.
11. J.L. Sudworth, private communication to E.J. Cairns, April 1981.
12. P.A. Nelson et al, Progress Report for the Period October 1977 - September 1978, Argonne National Laboratory Report ANL 78-94, November 1978.
13. H. Shimotake et al, in Proceedings of the 11th IECEC, Volume 1, American Institute of Chemical Engineers, New York, 1976, p. 473.
14. S.M. Zivi, in Annual DOE Review of the Lithium/Metal Sulfide Battery Program, June 1979.
15. R. Bellows et al, in Extended Abstracts of the Electrochemical Society Meeting, Los Angeles, October 1979, 79-2, Abstract 112.
16. E.J. Cairns and E.H. Hietbrink, Electrochemical Power for Transportation, in Comprehensive Treatise of Electrochemistry, Volume 3, J. O'M Bockris, B.E. Conway, E. Yeager, and R.E. White, eds., Plenum Publishing Corporation, 1981, p. 421.

17. E.J. Cairns, in Materials for Advanced Batteries, D.W. Murphy, J. Broadhead, and B.C.H. Steele, eds., Plenum Publishing Corporation, New York, 1981.
18. E.J. Zeitner and J.S. Dunning, in Proceedings of the 13th IECEC, Society of Automotive Engineers, Warrendale, PA, 1978, p. 697.

This work was supported by the U.S. Department of Energy under Contract W-7405-ENG-48.

The Chemical Industry: Future Energy and Feedstock Requirements

By P. G. Caudle
CHEMICAL INDUSTRIES ASSOCIATION LTD., 93 ALBERT EMBANKMENT, LONDON SEI 7TU, U.K.

Introduction

A year ago, a paper on this subject could have readily been based on a number of similar studies over the past two or three years [1, 2, 3], suitably updated with the latest statistics, and with degrees of emphasis on well recognised technical and economic developments dependent on personal views regarding the speed with which these would play a significant part in future trends.

It is clear that such an approach is no longer valid. The development of the OECD economies in the post-war period was already subject through the '70s to constraints which can be labelled "limits to growth"; on these have been superimposed two massive increases in oil prices, in 1973/4 and 1979/80. While the first such price rise was accommodated fairly readily in macro-economic, financial and commercial terms, the second may have triggered a new structural situation within the Western economies, to which we have yet fully to adjust. My present paper will consider the possibility that such adjustments over the next decade will involve more than just a moderate scaling down of previously forecast growth rates, and more than a marginal "moving over" to accommodate a slow evolution of production of petrochemicals and other energy intensive materials from outside the OECD area. To forecast energy and raw material requirements on a "no change" basis would do no more than perpetuate the over-optimism of which many of us - myself included - have been guilty during the '70s. It is time to allow for pessimism about the "natural" growth and development of the Western economies in general and their chemical industries in particular, and to look hard at what must be done to ensure that we still have viable manufacturing and chemical industries by the end of the '80s.

These will not be the same sort of industries as at present. In respect of growth of energy and feedstock requirements, the historic chemical industry pattern will change considerably, and any forecasts must look hard at possible structural changes in our industry, both nationally and internationally, as well as the likely growth rates which may be achieved within the economies served by the chemical industry.

Although my own and other quantitative forecasts of the last few years must be discarded, my 1978 paper [3] did anticipate these future uncertainties. May I therefore take as my "text" for today's presentation, that paper's penultimate paragraph:

> "My basic conclusions appear dull and conventional - there will be petrochemicals for a long time to come, with a gradual replacement by raw materials and process routes, most of which were already familiar in the early half of the century. In fact, the future is far from dull, if only because it is so uncertain. Instead of a steady and rapid expansion of production and techniques based on a single feedstock in plentiful supply, the chemical industry has to grope towards a variety of new sources of raw materials and energy at a pace made undertain both by supply and by demand factors. It has to base its future development on uncertain capital costs and even more uncertain raw material and energy costs. It has to take account of a host of social, political, geographical, and logistical factors which it could hitherto almost ignore. And it has to tackle - if my analysis is correct - botanical and biochemical technologies as well as those based on chemistry and physics."

Where do we currently stand?

Before speculating about future developments, one needs as a datum the energy and raw material use by the chemical industry, set in the context of overall economic activity in the corresponding regions in which it operates.

The regions most relevant to our discussion are those which are conveniently labelled "OECD". There are good reasons for this, not least that fairly reliable statistics are available! The more pertinent reason is that such regions broadly share the same social and economic objectives and operate, in theory at least, according to similar economic rule-books. The impacts of oil-rich "Third World" nations, the Communist nations and the poorer parts of the "Third World" can, in economic jargon, be regarded as "exogenous" and considered as a separate issue.

Within the OECD, one can distinguish a number of distinct situations, ranging from an energy-rich Norway to an energy-impoverished Japan. I shall, however, concentrate primarily on the EEC (and, within the EEC, the UK), with reference to factors which create non-typical situations within or directly affecting this area.

Figure 1 shows the 1979 chemical industry turnover in the four broad zones of the OECD.

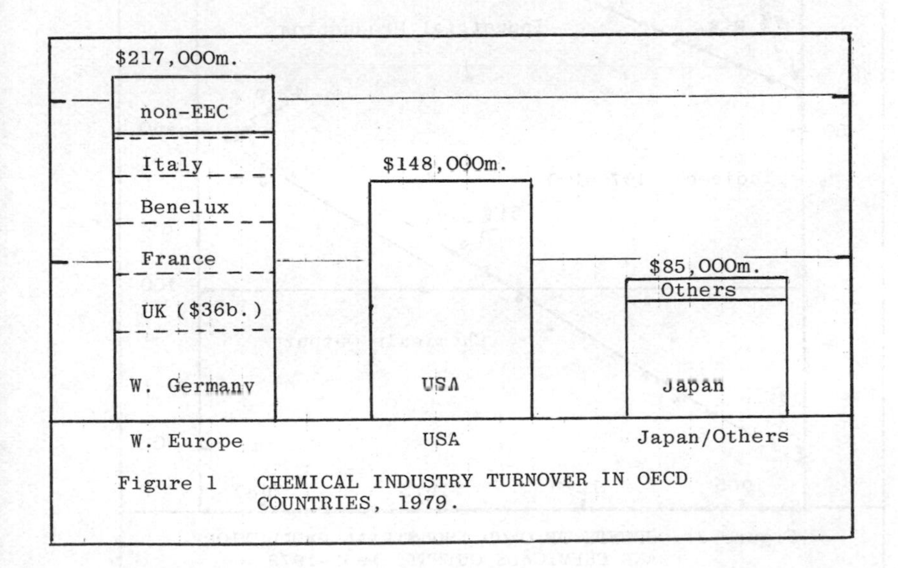

Figure 1 CHEMICAL INDUSTRY TURNOVER IN OECD COUNTRIES, 1979.

The total OECD chemical 1979 output, around $450 bn., represents about 65% of world output in that year, the remainder being split 25% in Communist countries and 10% in the "Third World".

1979 was a reasonably prosperous year for large parts of the chemical industry. Figure 2 shows the trend within the OECD up to that year, with growth rates in the '70s not as high as in the two previous decades, but still averaging 6-7% p.a. 1980 has proved to be quite another story, but for the moment I would like to stay with 1979, and turn to the energy and feedstock inputs needed to sustain the outputs shown in Figure 1. In terms of quantities, these are shown as m. GJ in Figure 3.

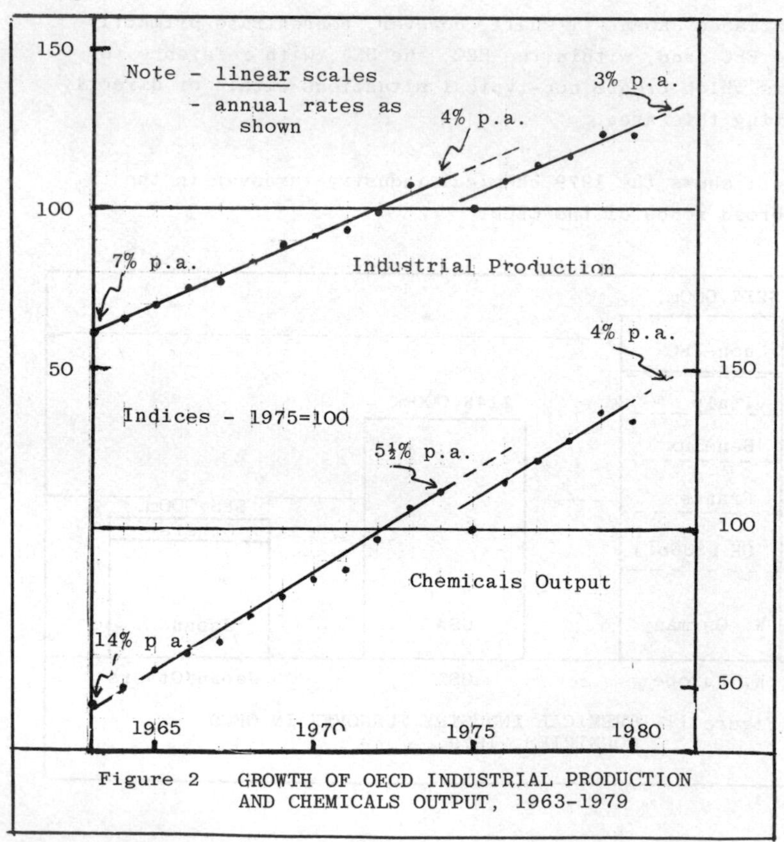

Figure 2 GROWTH OF OECD INDUSTRIAL PRODUCTION AND CHEMICALS OUTPUT, 1963-1979

Factors to convert the quantities in Figure 3 to value terms will obviously be different in each main region, and within regions. In the UK, we have calculated a total "energy bill" in 1979 of about £830m. (6% of turnover) and feedstocks probably account for a further £800m. making a total of £1630m. (11% gross turnover). The proportions for W. Europe were probably not dissimilar in 1979. (It should be noted that gross turnover includes sales <u>within</u> the chemical industry, and as a proportion of <u>external</u> sales, energy and feedstocks probably accounted for about 14% of turnover in the UK in 1979.)

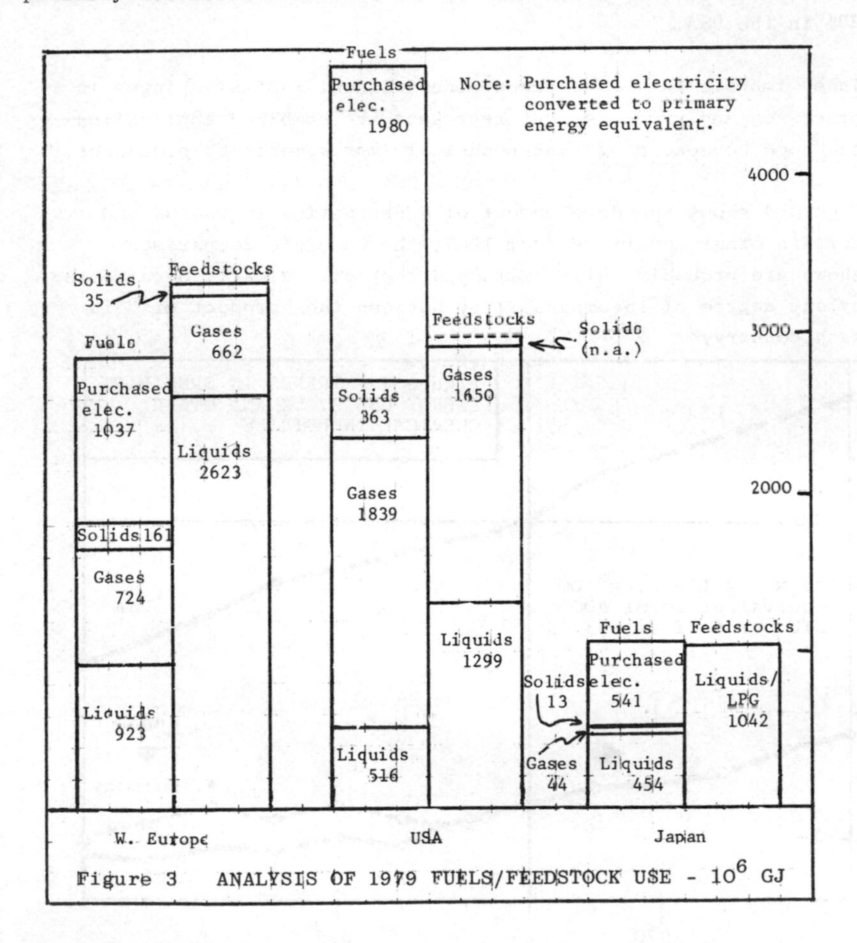

Figure 3 ANALYSIS OF 1979 FUELS/FEEDSTOCK USE - 10^6 GJ

The upward trend of energy use has not been as rapid as that of output, the improvement of efficiency having been a feature of technical and commercial developments through the post-war era rather than a direct response to the energy price rises of 1973 and after.

Put another way, the specific energy use index in the UK fell from 100 (1970) to 81 (1979), an improvement in efficiency of use of 19%. Similar improvements have been achieved in other OECD countries - i.e. 18% in W. Germany 24% in France, and 22% in the USA.

These improvements have taken place without radical changes in processes and products, but represent the combined application of "good housekeeping" and technical improvements to processes.

Figure 4 shows the development of conservation in the UK and certain other countries from 1967; the absolute comparisons shown are probably valid, although they will clearly be affected by any degree of incomparability between the "product mix" in each country.

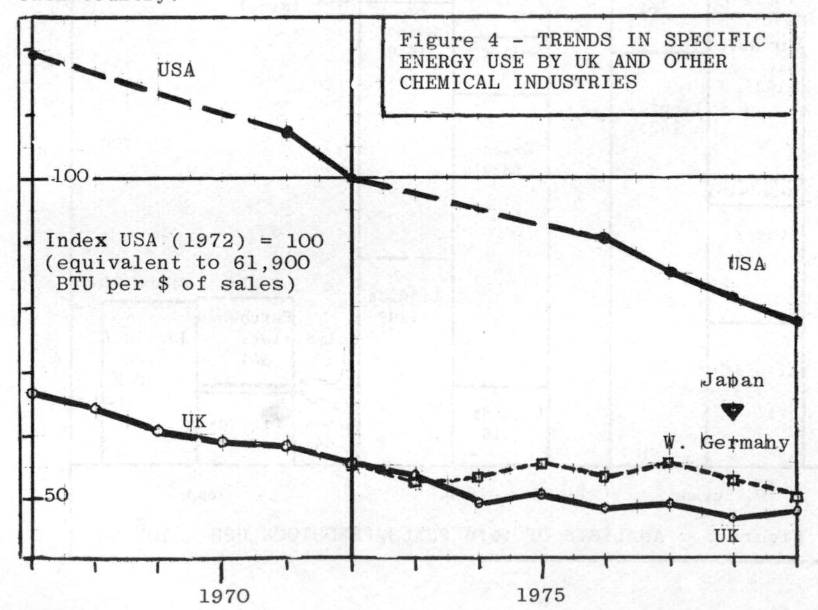

Figure 4 - TRENDS IN SPECIFIC ENERGY USE BY UK AND OTHER CHEMICAL INDUSTRIES

1980 and Beyond

For the UK chemical industry, 1980 represents a break in trend quite distinct from previous recession years, such as in 1971 and 1975. This is apparent from Figure 5, from which it is also clear that in other parts of Europe, the plunge has been less steep and, unlike in the UK, there are some genuine signs of recovery of output.

For the EEC and W. Europe as a whole, projections of output, energy and raw material use made by CEFIC[4] (Conseil Européen des Fédérations de l'Industrie Chimique) during the last few months are based on two "scenarios" - "optimistic" and "low growth". Translated into energy consumption terms, and allowing for an annual reduction of 1% in specific energy consumption over the period 1979-89, leads to the extrapolation shown in Table 1.

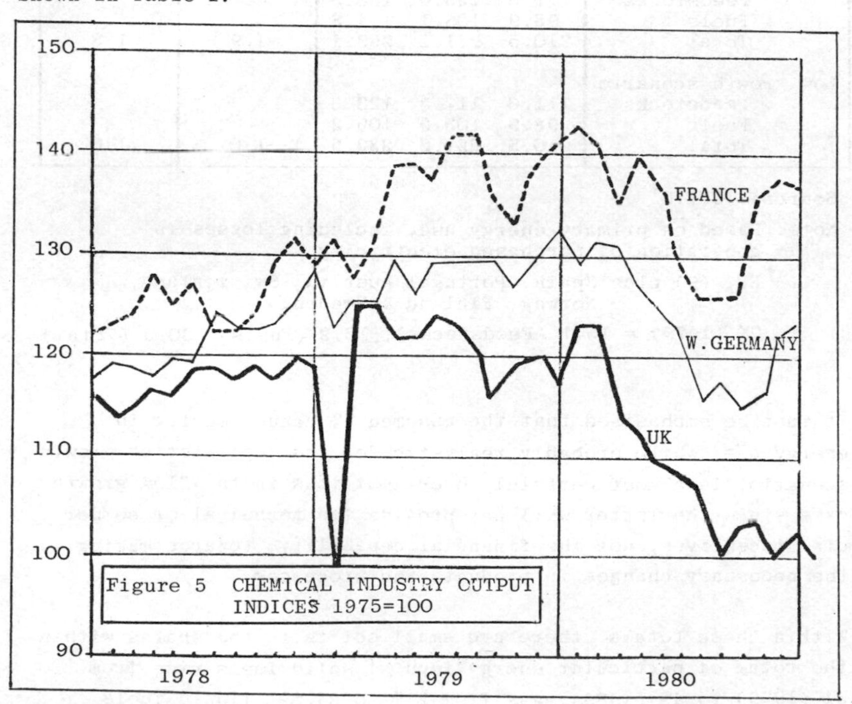

Figure 5 CHEMICAL INDUSTRY OUTPUT INDICES 1975=100

Table 1

ENERGY REQUIREMENTS OF THE W. EUROPEAN CHEMICAL INDUSTRY

Million tons coal equivalent (1 TCE = 29.31 GJ)	1979	1984	1989	Growth Rate % p.a. 1979-1984	1984-1989
EEC region ("9")					
Optimistic scenario					
Feedstocks	99.9	110.1	120.9		
Fuels	82.9	87.4	94.4		
Total	182.8	197.5	215.5	1.6	1.7
Low growth scenario					
Feedstocks	99.9	105.0	110.4		
Fuels	82.9	87.1	91.5		
Total	182.8	192.1	201.9	1.0	1.0
W. Europe†					
Optimistic scenario					
Feedstocks	111.6	125.0	138.3		
Fuels	98.9	106.1	114.8		
Total	210.5	231.1	253.1	1.9	1.8
Low growth scenario					
Feedstocks	111.6	117.3	123.3		
Fuels	98.9	103.9	109.2		
Total	210.5	221.2	232.5	1.0	1.0

Source: CEFIC[4]

Note: Based on primary energy use, including losses in generation of purchased electricity.

†EEC (9) plus Spain, Portugal, Austria, Switzerland, Norway, Finland & Sweden.

UK (1979) = 15.1 (Feedstocks); 15.2 (Fuels); 30.3 (Total)

It must be emphasised that the assumed 1% annual saving in energy use, while probably realistic for the "optimistic" growth scenario, is almost certainly over-ambitious in the "low growth" case since the latter will not provide the technical or commercial incentives, nor the financial capability, towards making the necessary changes in products and processes.

Within these totals, there are small shifts in the shares within the total of particular energy inputs: solid fuels rise from 3% (1979) to 4% (1989); gas from 23% to 24.5%; liquid fuels fall from 57% to 54.5%; and electricity remains constant at 17%. UK (1979) proportions were 2%, 25.5%, 54% and 18.5% respectively.

The requirements of energy containing raw materials commensurate with the "optimistic case" are detailed up to 1984 in other recent CEFIC surveys [5,6]. These have been criticised as over-optimistic, and in its energy survey[4] CEFIC has now provided a "low growth" estimate for feedstock use, extrapolated to 1989. The overall requirements of both energy and energy containing feedstocks under both scenarios are summarised in Table 1.

While the increase in total energy/feedstock inputs into the European chemical industry over the decade may well be within the potential ability of supply from the various indigenous and imported energy sources, mention must be made of the growth of naphtha requirements. Even allowing for a sizeable increase in the use of alternative feedstocks (C_2/C_4 gases, gas oil etc.) in certain plants - particularly those in the UK - forecast European naphtha requirements rise from a 1979 level of 51m. tons to a 1989 level of about 57m. tons, at a time when there is likely to be an increasing pressure on the "light end of the barrel" as a result of reduced crude runs and sustained demand for gasoline and other "white" products. Clearly, this pressure will provide a strong incentive for chemical manufacturers to accelerate the shift towards C_2/C_4 feedstocks and/or to purchase "ready-made" naphtha derivatives from outside W. Europe, whilst in parallel the refiners will no doubt press on with existing plans to install conversion equipment designed to lighten the product barrel.

We must now revert to the exceptional UK position, as illustrated in Figure 5. Firstly, it should be emphasised that within the UK chemical industry, the collapse of demand and output shown at January 1981 will NOT be rapidly reversed; indeed, current expectations following the recent Budget allow for, at best, a plateau of output at around the present level for the next two years - that is, at a level about 20% lower than in 1979 which was in no sense an exceptional year.

The cause of the UK calamity - it is no less - in regard to manufacturing industry generally can be ascribed, dependent on ones political and economic stance, to the culmination of long-developing defects in the UK economy, or to the incidence of ill-judged economic measures since 1979. The longer term aspects of the problem clearly include:

- poorer productivity, management and enterprise within UK manufacturing industry generally.
- delays to membership of the EEC, and consequential unreadiness to "stand the heat of the EEC kitchen" - or to enjoy the fruits on the EEC table.
- industrial problems affecting downstream customer industries, energy, transport and plant construction.
- social attitudes, favouring non-industrial endeavours and objectives, and leading to expectations of increased material welfare not justified by material achievement.

Shorter term factors include:

- depression of domestic demand by recent and current economic and fiscal policies.
- high interest rates, employment surcharges, local rates etc.
- the effect of N. Sea oil (and some of the above factors) on the value of sterling, with consequential adverse effects on export margins, import volume and home sales of products.
- the very rapid increase in costs of services (in particular energy) provided by State monopolies.

Most of these factors, both short and long term, bear directly on the UK chemical industry. All of them affect the performance of our UK customers, and even if our exports can be maintained at or above the present (40%) share of chemicals production, it is essential that there should be a recovery of a strong UK customer base for UK chemicals output to the extent this is possible in the context of recent irreversible changes. Unless many of the above adverse factors are rapidly eliminated or ameliorated, it is self-evident that the UK chemical industry will be hard-pressed to maintain its low 1980 output, let alone participate in the recovery expected in the rest of Europe during 1981/82.

Having regard to the above, the UK contribution to the energy and feedstock figures shown in Table 1 must be considered highly provisional and even suspect, and it cannot be assumed that the 1979 share of EEC energy use (16.6%) will be maintained.

Special Factors in the Medium Term

There are two broad ways to tackle the problems the world now faces regarding energy constraints: to assume that "market forces" will provide the optimum solution, the "lying back and enjoying it" approach; or the alternative assumption that one should attempt to control ones own destiny. In respect to finding and producing oil and gas, the UK has happily taken the second course. In virtually all other respects it has proceeded down the first road. Those who need energy and feedstocks are certainly not enjoying the result, since most of our competitors try and succeed in creating a technical and economic environment much more favourable to their chemical industries than those provided by "market forces" in the energy field. "Market forces" have been described by Prof. James Meade as "the worst mechanism there is - apart from all the others", but this qualification seems hardly valid in a scene dominated by international cartels (OPEC) and or own State monopolies. As an alternative, our own Government offers "economic pricing principles" - giving, as recently suggested by a BBC commentator, the "prices the bureaucrats in Whitehall want to see". It is not clear whether these principles should be oriented towards long run marginal production costs, long run marginal supply prices, market clearing costs, resource costs, historic accounting costs, inflation accounting costs, opportunity costs (and so on!) - all come within the corpus of valid economic theory. Each and every one of these have recently been adduced to justify the present peculiar and disadvantageous UK price structures for industrial energy sources, as identified in the recent NEDC Task Force report.[7]

Alongside such domestic confusions, we find that other competing nations, lacking indigenous oil and gas, have developed long term supplies of these from abroad at quite favourable prices. They have, moreover, successfully developed their own "non-exportable" resources - brown coal, hydroelectric and nuclear power for example, and, unlike the UK, do not regard these resources as milch cows for central government revenues. Future developments abroad based on shale, tar sands and a rapid expansion of nuclear capability seem certain, and in parallel, both Japan and continental Europe are increasing their docks, transport and technical potential for handling low cost imported coal, which is likely to become increasingly important as an energy source, in both conventional and fluidised bed boilers.

Up to this point, the difference between the UK situation and that in most other competing countries does not depend on subsidies or "unfair" practices. But in addition, many countries see a necessity, in terms of national strategy, to retain an effective base of energy and feedstock-intensive manufacturing industry, thus justifying deliberate intervention in the pricing and financing of the inputs to such industries. Given the increasing dominance and dubious stability of many parts of the world where fossil fuels are inherently cheap to produce (in particular, the OPEC states), it is far from obvious that such intervention, at a national strategic level, is "unfair" or reprehensible.

Regardless, however, of economic theories or political judgments, the end result in 1981 is an international comparison between principal energy sources which places UK prices at levels 10-25% higher than in the rest of the EEC for fuel oils and gas, and up to 50% higher for supplies of electricity. In turn, Continental prices are much higher than those which apply in the USA and Canada, and certain other parts of W. Europe such as Norway and Spain.

Feedstock prices tend to be specific to particular contracts and locations, the major distortions at the present time arising from the continuing regulation of most US gas prices at far below world levels (affecting methane directly and C_2/C_4 hydrocarbons indirectly) and the policy decision in Norway to "decouple" gaseous feedstocks from world market prices.

Despite statements to the contrary, Continental EEC energy price levels are not rising faster than in the UK, and despite US decontrol of oil, the currently proposed phasing of their regulation of gas will not bring their prices into line, even by 1987. In particular, the French nuclear programme - designed to move the French energy economy dramatically towards this basis over the next decade - is already providing very cheap electricity and has recently been projected to provide power, at the end of the decade, at 13 centimes/kWh (1.2p/kWh) in terms of 1980 money. Additionally, there will remain very considerable pressures from OPEC states (in the Middle East and Mexico in particular) to set up petrochemical and other energy intensive industries based on "local" costs of energy and feedstocks, and there is no sign that Norway or Canada will shift from their present policy of using "local" prices for oil, gas and hydroelectric power.

It is, I believe, not at all irrelevant to introduce these international politico-economic factors into a paper presented mainly to a body of UK chemists. If, as chemists, you are interested in the impact of your professional achievements and expertise on UK industrial performance, it is - perhaps regrettably - the incidence of such factors over the next decade which will be much more important than the developments in your laboratories and pilot plants.

Unless, in the UK, there is a change of national policy towards energy and feedstock pricing, it is unlikely that we will see any significant use being made of the positive factors, in particular those relating to security of supply, which could be provided by UK oil and gas. Thus, in relation to the most basic of petrochemicals, ethylene, recent unpublished studies have identified a range of projections as in Figure 6.

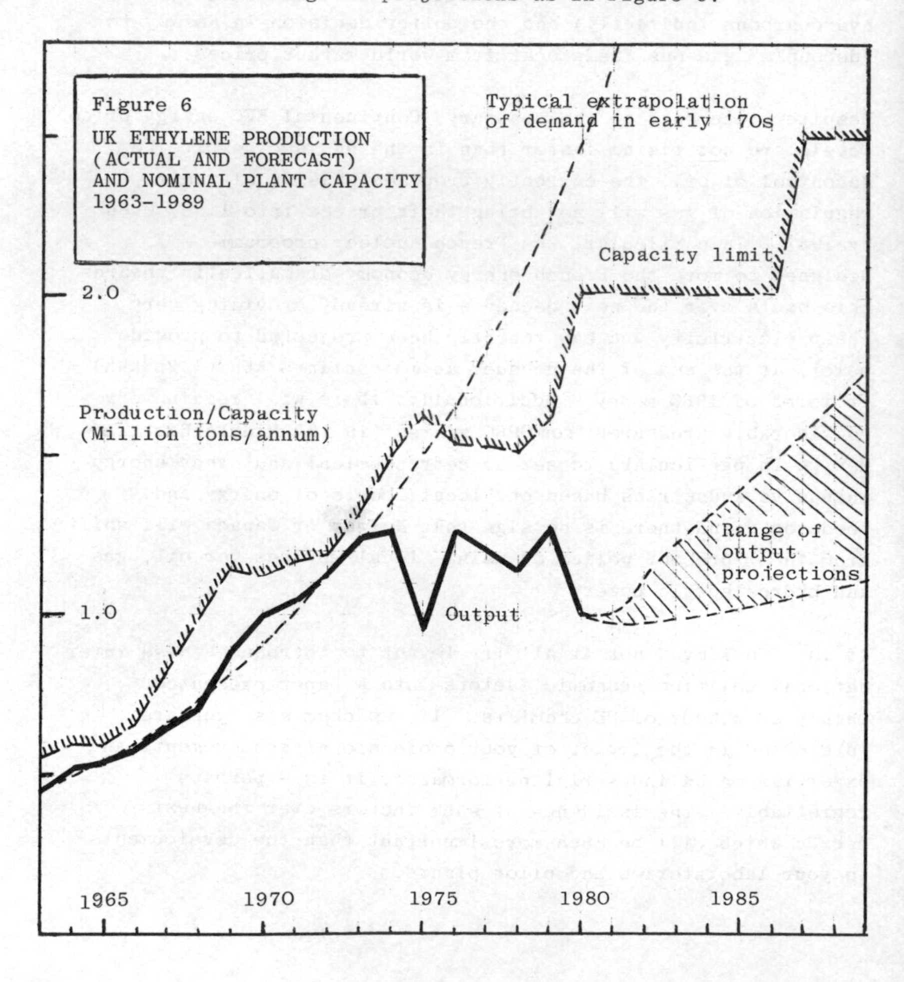

The "low" projection assumes no special advantage in terms of feedstock costs compared with "world" naphtha prices or, if C_2/C_4 gases are used, the present BGC pricing structure for natural gas as supplied to the industrial energy market at 25-30p/therm in 1980 money terms. In such circumstances, there is no reason for the UK to achieve even a retention of market share of olefin production within the EEC/W. European area. Even given a much more favourable feedstock price basis which might lead to the "high" projection, it is noteworthy that UK ethylene production remains well below the capacity of plants available or under construction.

"Special factors" can, in other areas, be much more favourable to increased market share than, under present policies, obtain in the UK. Thus, current US pricing of ethane is a factor in recent market pressures in Europe on its derivatives - e.g. vinyl acetate. Chlorine and hydrocarbon/chlorine derivatives are being produced in Norway much more cheaply than in the rest of Europe. UNIDO, UNCTAD and "Euro-Arab" channels are being directed at securing entry into Europe of petrochemicals based on cheap hydrocarbon and energy inputs, with the added pressure arising from the need of many European countries to secure oil supplies as a "package deal". "Buy-back" deals with E. Europe have been established which provide commodity chemicals as the payback for technology.

Taken together, these factors will represent a curtailment of <u>production</u> in W. Europe compared with that implicit in Table 1, even though the <u>demand</u> for petrochemicals and polymers may be at the level assumed. Similar reductions may apply to other energy-intensive products of the chemical industry (inorganic chlorine derivatives, metal oxides etc.) where these can readily be made in, and transported from, low energy cost regions.

In summary, the energy and feedstock projections outlined in Table 1 probably represent the maximum indigenous requirements of these inputs over the next decade. Below the "low growth" scenario lies a "de-industrialisation" projection down which, under present policies, we have advanced some distance in the UK during the past two years. For the next decade, it is not a case of forecasting how far and fast we go down this road, but of deciding, as an act of national policy, whether this is the road we want to be on; if we do not, decisions must be taken to change course at the level of national energy strategies.

Long Term Trends and Strategies

At this point, I will leave the distortions induced by present political and economic strategies and begin once again to look at how, given the chance, the world of science, engineering and business can help us find economic solutions to our energy and feedstock problems.

In my earlier paper[3], I tried to identify the most important new or "revived" routes to energy-intensive chemical products, as in Figure 7.

Some of these routes have local importance for strategic reasons (e.g. the Fischer-Tropsch operation in S. Africa) or because they can use "low technology" and are dependent on the availability of local resources (e.g. ethanol in Brazil). Others can probably be excluded from early consideration for reasons of technical economics or practicality - coal hydrogenation to aromatics, and phototropic production of hydrogen are in this category. My own choice of world scale "winners", at least during this century, narrows down to two candidates:

- production of specialised biologically derived chemicals which do not involve serious degradation of the original material and the consequent need to re-synthesise carbon chains;
- production and conversion of methanol.

Figure 7 ROUTES TO ENERGY-INTENSIVE CHEMICAL PRODUCTS

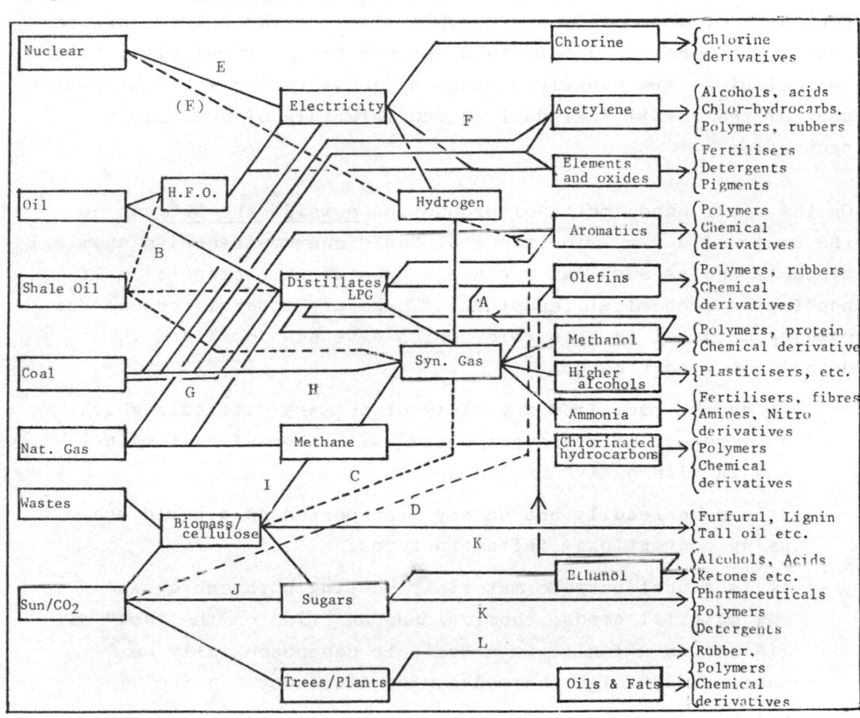

Key - New Routes

A. The conversion of methanol to liquid and gaseous hydrocarbons.
B. The use of shale oil as a substitute for crude oil and/or coal.
C. The aerobic conversion of biomass to synthesis gas.
D. The phototropic production of hydrogen in biological systems.

- Old Routes with increased potential

E. Production of hydrogen by electrolysis (or direct nuclear heat).
F. Production of acetylene via carbide or a plasma reactor.
G. Coal hydrogenation (requires hydrogen or syngas).
H. Coal gasification (with or without Fischer-Tropsch synthesis).
I. Anaerobic fermentation of natural and/or waste materials.
J. Increased production of sugars either naturally, or by acid or enzymatic hydrolysis of cellulose; and subsequent fermentation or conversion.
K. New products (e.g. lignin) from existing natural materials.
L. Increased yields from existing, new or modified species.

The first of these categories represents products which are, in general, not especially energy-intensive; nevertheless, natural rubber, terpenes, and materials derived from natural oils, fats and cellulose are clearly capable of substituting for a sizeable part of the fossil fuel derived raw materials of equivalent properties.

On the other hand, methanol production outside the EEC can be the basis of a very wide range of basic energy-intensive chemical products and as such can replace a considerable proportion of naphtha as a chemical feedstock. There are several reasons for backing methanol in this role, which have been outlined in detail in recent papers[8,9,10]:

- it can be made from a variety of primary materials which, as a result of transport costs, will themselves remain low in alternative use value;
- it can be readily and safely transported on a world scale using conventional infrastructures;
- it is a dual purpose material, meeting both energy and raw material needs; chemical use can "ride on the back" of its use as a fuel; as a fuel, it can potentially be used at very high thermodynamic efficiency (in fuel cells);
- production technology is well established, but is also capable of further economic advances.

Set against these positive factors are its rather high cost of transport compared with LNG (if conversion costs from gas are included); its toxic and mildly corrosive properties; and the unreadiness of existing energy markets to use it as a fuel.

Despite these reservations, I believe it is the energy and feedstock source we shall see most rapidly developed over the next two or three decades.

The first positive factor is of great strategic significance. Europe will not want to see itself exchanging dependence on one cartel for another, but it should be able to look forward to methanol supplies not only from low value OPEC gases (which may otherwise be flared) but subsequently from a number of world sources of low cost coal/lignite and, somewhat later, shales and tar sands. This diversity should generate a reasonable level of world competition and ensure strategically reliable and economic European supplies.

Unlike the movement of LNG, methanol transport presents no serious safety hazards, and much of the existing world transport infrastructure can be adapted to methanol transport; if necessary the final upgrading to "feedstock" purity can be applied to fuel grade product at the point of use, minimising the use of specialised transport facilities.

Of the various technical steps needed to produce methanol and use it effectively as a chemical raw material, the only process not yet commercialised is the conversion step into hydrocarbons, (which can then be used in conventional chemical pathways and energy markets); this step is now being installed as part of the N. Zealand scheme to convert natural gases to liquid hydrocarbons; successful commercialisation of this operation should provide an important stage in the adoption of methanol as a major petrochemical feedstock.

I have not attempted to set out the likely economics of methanol-based petrochemicals; these will depend very much on raw material valuations,scale, financing basis and distance of transport. Two things seem fairly certain at the moment. Firstly methanol produced on a very large scale at locations where primary raw materials are of low cost and (because of distance from markets,of low alternative use valuation) can already compete as a fuel with conventional hydrocarbons. Secondly, it will be quite uncompetitive to produce methanol in Europe from indigenous or imported primary raw materials (such as coal) costed at "European" energy values, in comparison with production overseas adjacent to the source of raw materials.

Conclusions

In summary, the likely development in the short and medium term of energy inputs into the UK and European chemical industry will be at a much more moderate rate than in the past; the UK position, far from expanding rapidly as a result of our N. Sea resources, may remain severely contracted unless there is a significant shift in national energy and industrial strategies towards energy-intensive and other "heavy" sectors of manufacturing industry.

In the longer term, the European chemical industry must take steps over the next few years to participate in the establishment of new sources of primary fossil fuels in distant locations and to contribute to the technological and commercial development of these for production of its principal basic chemicals. Alongside these developments, it must use its own technological and research expertise to advance the production of chemicals based, without degradation of molecular structures, on naturally occuring products, many of which may be most economically produced in "Third World" countries. Finally, it must keep up with other high-technology regions (e.g. N. America and Japan) in developing the chemical technology of high value products (including especially those derived from sophisticated biotechnology) which are less susceptible to the burden of increasing energy costs.

All three of these approaches must be pursued if European chemicals are not to contract and become dangerously dependent on imported petrochemical intermediates and commodity polymers. Unfortunately, many of the major decisions needed are essentially political rather than technical or commercial. Whether we like it or not, the future of the UK and European chemical industry over the next several decades thus depends not on chemists, engineers and entrepreneurs but on politicians and bureaucrats in London, Brussels and other capital cities. Clearly, we must do our utmost to ensure that their decisions are well informed, and in the best <u>long-term</u> interests of the community in which we all live.

Acknowledgement

The author gratefully acknowledges the very considerable contribution of his colleagues in CIA and CEFIC to ideas and data in this paper. However, the views expressed are those of the author personally and should not be attributed to the above organisations or their constituent members.

References

1. R. Pennock, Chem. & Ind., 1978, 860
2. Shell International Chemical Co. Ltd., "Substitution Revisited" London, 1979
3. P. G. Caudle, FUTURES, 1978, 361.
4. CEFIC, "Energy Statistics - 1980", Brussels, 1981
5. CEFIC, "Survey on Olefins, 1974-1984", Brussels, 1980
6. CEFIC, "Survey on Aromatics, 1974-1984", Brussels, 1980
7. NEDC, "Report of the NEDC Energy Task Force", NEDO, London 1981
8. D. F. Othmer, Chemical Engineer, 1981, Jan., 19.
9. Papers in "Coal Chem - 2000", Institution of Chemical Engineers Symposium Series No.62, London 1981.
10. Papers in Hydrocarbon Processing, 1981, March, pp71-100.

Energy, Materials and Mankind

By W. O. Alexander
THE UNIVERSITY OF ASTON IN BIRMINGHAM, GOSTA GREEN, BIRMINGHAM B4 7ET, U.K.

Since we can assume that about half the world's energy utilised per annum is used in making metals and materials, I propose to outline some interrelating factors which will affect future materials strategy.

This view of tonnage materials for engineering and structural applications looks at metals, plastics, other non-metals such as concrete and timber, and evaluates likely developments following the recognition of limitations in availability of various energy and material sources and the properties of those materials.

Assessment of Total Energy Content of a Material

Publication of energy utilisation data in metallurgical and material processing operations has been widespread in recent years. However, such data is in rather an early stage of compilation. In some cases, there are discrepancies of x2 to x3 between values in different countries and between different firms. Recent work in the United Kingdom iron foundries, [1] and aluminium industry, [2] is also revealing that energy auditing in routine metallurgical establishments covering extended periods gives up to double the hitherto assumed values.

Difficulties have been experienced in assessing true works production data in total energy increments at each step, partly because metering of individual production units on a complex works site is not carried out and partly because the cost centres so dear to accountants seldom coincide with energy interests. Other errors stem from various assumptions as to the overall energy uses on a site, e.g. space heating, furnace standby losses, etc.[1,3,9] Partitioning of energy between products, say, of metal or plastics and surplus steam and electricity can also lead to disagreements between published values for energy requirements of even standard tonnage products.

The overall estimate of the total energy which has been already incurred by an ore as raw material or concentrate or virgin metal when it has arrived in the United Kingdom is also not readily calculated. Since so very little is now indigenous in the United Kingdom this factor is now important.

Another factor which incurs a good deal of controversy is the true fossil fuel factor to allow for generation of electrical energy. In the United Kingdom, allowing for electrical transmission and distribution losses, a multiplying factor of x4 is generous enough in favour of electricity. Losses are about 9%, 6% as line losses plus 3% for transformer switching and distribution losses, on a 28% thermal efficiency factor for the whole of power generation by the Central Electricity Generating Board.

Considerably more detailed work and agreed conventions will be necessary on an international basis before Total Energy data on materials is truly comparable. The Energy Audit Series of investigations by the Departments of Energy and Industry are probably the most accurate detailed reviews of energy usage in industry in the United Kingdom, but so far have only covered iron castings, building bricks, dairy products, bulk refractories, glass, pottery, brewing, aluminium and coke making. A recent publication gives the current agreed best practice in Europe for a range of plastics.[3]

Energy and Costs

It cannot be denied that market forces will ultimately predetermine the usage of energy and materials in whatever form, but this will only happen when the energy costs become a major proportion of the cost of any end product. In some cases energy is already 30% of the added value but, since the energy expenditures on incoming raw materials earlier in the processes are not quantified in separate financial terms, most managers and directors believe that the energy contents in their part of the process only represent 10 - 15% of their total cost. Rigorous energy auditing would reveal a much higher value and it has been suggested that total energy should now be considered as a fundamental unit which economists and accountants should use.[4] One could argue that energy and labour are almost the total cost of any metal or material, and it has been suggested that total energy should now be considered as a fundamental unit which economists and accountants should use.

In the interim, i.e. the next twenty years, it seems that total energy content in a product will not be accurately reflected in its cost. This is partly because the total energy content of a raw material is not known or appreciated, partly because various energy sources are so vastly different in basic costs,[5,6,] and hidden by state subsidies, and partly because energy auditing as yet cannot fairly apportion the energy used on a works site to each product group. There is also considerable discrepancy amongst observers in the field as to whether process energy varies significantly with output or not. Evidence within the occupied capacity 55 - 95% suggests that either connection may be valid.[7,8,9]

At a recent International Energy/Resources Conference in the United States of America, it was generally agreed that total selling prices of materials or products do not reflect their total energy contents.[10]

All these observations therefore justify the need to assess accurately and discuss the relevance of the total energy content of the major tonnage materials of the world.

<u>Unresolved Conventions for Total Energy Assessment</u>

The conventions for assessing total or gross energy seem to be fairly well accepted in Europe,[11,12] and I do not intend defining or elaborating them.

There are however, two other aspects of total energy auditing which will need general agreement and implementation. These relate to scrap recycling and the justification of manufacturing one life cycle only and/or dissipated materials.

The first is the value of energy which is to be assigned to recycled new and old scrap. This obviously consists of two components:

(i) The inherent total energy originally used to extract and make it the first time, i.e. as virgin metal or material, and

(ii) the 5 to 10% additional energy required to recycle into remanufacture.

Most firms would like to assess it at (ii) only, thereby gaining (i) for nothing. On the other hand, all scrap metals have an intrinsic financial worth which is fairly close to their raw metal values with the exception perhaps of iron and steel where it is artificially low. This financial value of scrap metal contains the total energy content of its original extraction and manufacture. It would obviously then be anomalous to ignore its total energy content when it is recycled, since thereby lower total energy values would be obtained for the new processed product. This view is further supported by the argument that all the energy utilised in initial manufacture is a world energy asset and must not be destroyed. The two references[11,12] suggest that energy credits must be given for recycled material.

The contrary view that on second and subsequent reprocessing much less additional or process energy is used per unit of product is attractive for short term marketing but cannot be tenable on strict energy auditing standards.

The second problem is that of composites. These I define for this particular problem as any geometric arrangement of two or more materials which cannot easily be separated into their component materials without excessive expenditure of energy or with low yields. This wide definition would cover such products as steel-cored tyres, glass-reinforced plastics, bi-metals, motor car radiators, honeycomb structures, laminates and other materials reinforced by various fibres or filaments such as carbon, boron, aluminium oxide, silicon carbide and certain organic fibres. In all such cases, their main reason for manufacture is that in service they contribute significantly to savings in operating energy or longer life.

Since by their very nature and manufacture they are unlikely to be recycled and are therefore lost, the energy used in making them must be justified. This can only be done by equating the energy used in their manufacture against the operating energy likely to be saved over the lifetime of the product. At the very lowest these two values should be equated, i.e. total energy used in complete manufacture to finished product = process operating or running energy saved over the average life of the product in service.

If the operating energy saved is greater than this so much the better. If it is less then the composite product must be justified for other valid reasons. Thus a composite or other single life cycle material might be justifiable:

1) if it gave the complete article a longer life;

2) the energy to make it was less than for a single material, but this is very unlikely;

3) a unique and more efficient energy technology was involved with no alternatives, e.g. micro-circuitry and solid state materials;

4) maintenance was significantly reduced.

It should be noted that this rigorous analysis will also be required for other single life cycle materials such as metals which are dissipated and plastics which are not recycled.

Again, present IFIAS convention would give no energy credit for waste and only enthalpy is included, "if the material is recycled".[11,12]

Consideration of the properties along with the total energy involved in manufacture give data which is absolute and likely to be fairly immutable, i.e. when the total energy contents have been more refined and generally accepted we have data on which future trends can be estimated with greater confidence than prognostications based on current costs and all the complications hidden and otherwise of subsidies, artificial internal transfer charges and monetary exchange values.

This thought is possibly of greater significance in the United Kingdom than in many other countries because our reserves are higher in energy and lower in metals and materials than most other countries. It is vital therefore that these energy sources are exploited for maximum long term benefit of the country and its inhabitants. Several countries already smelt aluminium with spare energy, some hydro-electric and some flare gas, and in this way export energy. Our energy is, however, far from cheap and would have either to make much more sophisticated products of low additional total energy or develop new routes for metals and materials which are much more efficient in energy utilisation.

Total Energy per Unit of Property

We must now consider the real value of the use of energy to make our metals and materials. Though we can evaluate the total energy consumed per kg of finished material, such information does not convey the inherent value of the product to prospective users. The determination of the value in total energy terms for the range of properties is of equal if not much greater relevance.

An outline of the properties of tensile strength, modulus of rigidity and fatigue strength for some common metals and materials is given in Table 1, together with their specific energy, i.e. total energy per kg of material, from which data the energy per unit of property can be readily calculated, as in columns 6 to 8, Table 1. The specific energy values for plastics use the European agreed best practices for the raw polymers to which an average 2.8 kWh/kg have been added for fabrication.[3] This addition makes the form of plastics equivalent to the metal semies such as strip, sheet, rod and angles.

As would be anticipated, total energy criteria throw a completely different light on the true values of some metals and materials to engineers and mankind.

For example, timber uses far less energy per unit of strength than any other material. Reinforced concrete is attractive at 145 - 250 kWh per meganewton of strength, followed by steels at 125 - 350, while cast irons give values of 73 - 292. It will be noted that the newer materials,

Table 1 Energy Consumption Related to Material Properties

Material	Tensile Strength MN/m^2	Modulus of Rigidity MN/m^2	Fatigue Strength MN/m^2	Density kg/m^3	Specific Energy kWh/kg	Total (kWh) energy per Meganewton unit of strength		
						Tensile Strength	Modulus of Rigidity	Fatigue Strength
CAST IRON Castings	400	45000	150	7300	4.0-16.0	73-292	0.65-2.8	194-776
STEELS								
EN1 free cutting rod	360	77000	193	7850	16.0	349	1.63	651
EN24 1.5 Ni/Cr0.25Mo rod	1000	77000	495	7830	16.0	125	1.63	253
Stainless 304 18Cr/8Ni sheet	510	86000	250	7900	32.0	229	2.94	487
NON-FERROUS METALS								
Brass 60Cu/40Zn sheet	400	37300	140	8360	27.0	565	6.05	1612
Aluminium sheet	300	26000	90	2700	79.0-83.0	711-747	8.2 -8.6	2370-2490
Duralumin sheet	500	26000	180	2700	79.0-83.0	427-449	8.2 -8.6	1185-1245
Magnesium alloy rod	190	17500	95	1700	115.0	1029	11.17	2058
Titanium " 6Al/4v rod	960	45000	450	4420	200	920	19.6	1964
PLASTICS								
Propathene GWM 22 sheet	35	1500	7.5	906	22.0	575	13.4	2660
Polythene L.D. XRM 21 "	13	84	3.25	920	22.0	1555	241.0	6225
Rigidex 2000 HDPE sheet	30	1380	4	950	28.0	932	21.5	7000
Nylon 66 A100 rod	86	2850	20	1140	45.0	595	18.0	1560
PVC (R) sheet	50	1680	12.5	1400	27.5	769	23.0	3080
REINFORCED CONCRETE	38	10000	23	2400	2.3- 4.0	145-253	0.55-0.96	240- 417
TIMBER Hardwood	14	4500	6	720	0.5	26	0.08	60
Softwood	5	2000	3	550	0.5	55	0.14	92
VITRIFIED CLAY	26	21400	-	2330	1.89	169	0.21	-
GLASS	100	30000	-	2500	3.30	83	0.28	-

aluminium, magnesium, plastics and titanium, all need greater total energy contents per units of strength. These range from 400 for duralumin to 700 for commercial aluminium, 1030 for magnesium, 920 for titanium and 500 to 1500 for the range of plastics.

Roughly the same ranking order for this range of materials is also true for modulus of rigidity and fatigue strength, though plastics are proportionately much more expensive in energy terms for these two properties than steels, cast iron or reinforced concrete.

Total Cost per Unit of Property

Table 2 considers the same materials and properties equated against cost instead of energy. The costs used are the average values of the basic prices in the semi-finished shape, i.e. as sheet/strip, rod, castings or mouldings, as quoted for tonnage quantities on 1st March, 1981. To give some idea of the conversion margin or added values between the raw materials and semi or fabricated products, Table 3 compares these prices. It must be noted that many of the costs are approximations and subject to market fluctuations. In the case of metals, the metal content value may be lower because 30 - 50% scrap metal may be included at a value usually lower than the virgin metal value.

For many of the materials this results in a somewhat similar ranking, i.e. the reinforced concrete, the steels and cast irons, followed by a range of polymers with aluminium alloys competing with plastics. However, there are a few drastic shifts in the ranking table: timber falls from cheapest in energy to almost dearest in cost terms. Two other marked shifts downward in ranking occur for stainless steel and titanium, possibly because of the high costs of melting, casting, working and getting good surface finishes. With the fall of these three materials to a high cost per unit of property, most of the other materials move up but maintain the relative ranking order.

This type of analysis can be extended to cover other properties of engineering and life performance. Furthermore, by a system of weighting and scaling the significance of a property in the overall performance

Table 2 Cost Related to Material Properties

Material	Tensile Strength MN/m^2	Modulus of Rigidity MN/m^2	Fatigue Strength MN/m^2	Density kg/m^3	Cost £/tonne	Cost (£) per Meganewton unit of strength: Tensile Strength	Modulus of Rigidity	Fatigue Strength
CAST IRON Castings	400	45000	150	7300	500	9.1	0.08	24.4
STEELS								
EN1 free cutting rod	360	77000	193	7850	250	5.5	0.03	10.0
EN24 1.5Ni/Cr0.25 Mo rod	1000	77000	495	7830	400	3.1	0.04	6.3
Stainless 304 18Cr/8Ni sheet	510	86000	250	7900	1250	19.5	0.12	4.0
NON-FERROUS METALS								
Brass 60Cu/40Zn sheet	400	37300	140	8360	1750	36.5	0.39	106
Aluminium alloy sheet	300	26000	90	2700	1100	10.0	0.11	33
Duralumin sheet	500	26000	180	2700	2100	11.3	22.0	31
Magnesium alloy rod	190	17500	95	1700	4000	35.8	0.39	72
Titanium " 6Al/4v rod	960	45000	450	4420	20000	92.0	1.95	196
PLASTICS								
Propathene GWM 22 sheet	35	1500	7.5	906	1400	362	0.84	169
Polythene L.D. XRM 21 "	13	84	3.25	920	2000	141	21.9	566
Rigidex 2000 HDPE sheet	30	1380	4	950	1700	59.6	1.30	403
Nylon 66 A100 rod	86	2850	20	1140	5250	69.6	2.10	300
PVC (R) sheet	50	1680	12.5	1400	1900	53.1	1.58	213
REINFORCED CONCRETE beam	38	10000	23	2400	30	1.9	0.007	3.5
TIMBER Hardwood	14	4500	6	720	500	24.0	0.07	56.1
Softwood	5	2000	3	550	300	33.0	0.08	55.0
VITRIFIED CLAY	26	21400	-	2330	80	7.2	0.008	-
GLASS	100	30000	-	2500	150	3.7	0.012	-

PRICES at 1.3.1981

Table 3 Raw Materials, Semi-fabricated and Conversion Margin Costs

Material	Costs (£) per tonne		
	Virgin Metal or Raw Material	Semi or Fabricated Form	Conversion Margin or Added Value
CAST IRON Castings	110	500	390
STEELS			
EN1 free cutting rod	110	250	140
EN24 1.5Ni/Cr0.25 Mo rod	100	400	300
Stainless 304 18Cr/8Ni sheet	400	1250	650
NON-FERROUS METALS			
Brass 60Cu/40Zn sheet	620	1750	1130
Aluminium alloy sheet	810	1100	290
Duralumin sheet	950	2100	1150
Magnesium alloy rod	1370	4000	2600
Titanium alloy 6Al/4v rod	3500	20000	16500
PLASTICS			
Propathene GWM 22 sheet	500	1400	900
Polythene L.D. XRM 21 sheet	540	2000	1460
Rigidex 2000 HDPE sheet	600	1700	1100
Nylon 66 A100 rod	1920	5250	3330
PVC (R) sheet	350	1900	1650
REINFORCED CONCRETE beam	25	30	5
TIMBER Hardwood		500	
Softwood		300	
VITRIFIED CLAY		80	
GLASS		150	

PRICES at 1.3.1981

sense, it is possible to determine the cheapest material in total energy or in cost terms for any predetermined combination of properties. Such estimates can be readily carried out by computer.[13]

Future Trends

In this brief review I am more concerned with the total energy : metal/material interface and the repercussions of this consideration. As a general indication of present thought on the extraction side I can hardly do better than bring out some observations from Professor H. H. Kellogg in his Julius Wernher Memorial Lecture.[14] This highlights the fact that true conservation saves energy and capital and the environment. He also illustrates the relative inefficiency of using electricity as compared with fossil fuel in many metallurgical processes with the possible exception of electrolytic processes. This is also true of the capital cost of the different energy sources in $/GJ/year. Similarly, hydro-metallurgical processes do not look so attractive as pyro-metallurgical routes and particularly so where tonnage oxygen can accelerate the process, increase productivity and yield and conserve capital, thereby saving 30 - 40% of energy. Yet another important aspect which he touches on under the heading of industrial symbiosis is the cogeneration of steam and electric power, waste heat recovery and environmental dust and fume control. All of which have been achieved in the past in various isolated cases across the world but which from now on with the increasing costs of power will merit more profound examination particularly in the location of new industrial extraction processes.

Availability - Energy

Another very serious factor to bear in mind is that as the resource becomes more difficult to discover and win, so the total energy required per tonne of finished product rises. This is well exemplified in the case of lead, Figure 1,[15] which indicates that below a certain lead content in an ore body, the energy cost becomes prohibitive and so lead supplies would decrease. It is recognised of course that this is a generalisation and that other factors also operate to affect the total energy used and costs. These may be size of mine and location, throughput of ore, overburden and waste as well as proportion of profitable

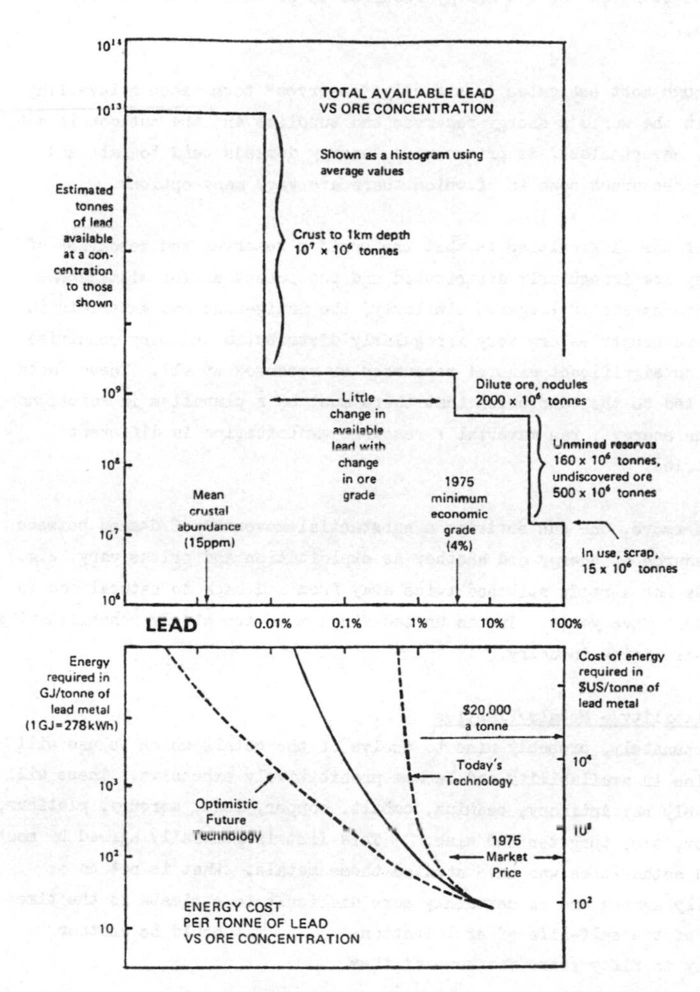

Figure 1 Availability and energy cost of lead vs ore grade

by-products. One saving grace to this difficulty which could considerably counterbalance it would be to recycle the metal already extracted. This can be and is done with the expenditure of one-tenth to one-twentieth of the energy required to produce it in the first place.

Although most estimates in strictly "reserves" terms show a levelling off in the world's energy reserves and supplies and the outlook is not good, nevertheless, as prices rise, energy demands tend to fall and other resources come in of which there are very many options.

One of the difficulties is that the world's reserves and resources of energy are irregularly distributed and the potential for mismanagement by Governments is larger. Similarly, the indigenous raw materials in various countries are very irregularly distributed and many countries have no significant mineral resources or reserves at all. These facts have led to the conclusion that there must be a pluralism in solutions in the energy : raw material : resource exploitation in different countries.

Furthermore, one can envisage a substantial movement of demand between one source of energy and another as exploitation and prices vary, e.g. Canada has already switched twice away from and back to natural gas in the last five years. In the United Kingdom we are slowly rehabilitating our coal mining industry.

Availability - Metals/Plastics

Unfortunately, probably nine to twelve of the metals which we use will decline in availability and become prohibitively expensive. These will probably be: Antimony, cadmium, cobalt, copper, lead, mercury, platinum, silver, tin, tungsten and zinc.[15] This list is generally agreed by most world authorities who have studied these metals. What is not so readily agreed and is certainly more difficult to estimate is the time span of the half-life of exploitation; it could be another thirty to fifty years for some of them.

Fortunately, the remaining metals and non-metallic materials are much more abundant and need not worry us so far as availability is concerned. What will be a very significant controlling factor will be the energy resource which we must assume will also have to be rationed or apportioned about the year 2000, and beyond. In many countries, this could well be earlier, but fortunately, in the United Kingdom we may be energy fat until beyond that date unless we have to export our oil and coal to trade for materials imports. The fact that oil is both the feedstock and the main energy source for most plastics means that their prices and whole economics relative to many metals and other materials will undergo a radical change.

Longer Life and Better Recycling is Essential

One immediate and relatively easy way to conserve energy is to aim to double the life of all products and then double them again. There is no doubt that many current metals and materials will slowly level out in growth due to a variety of circumstances. The reasons are lower grades of ores only available and the increasing cost of energy to mine and extract these ores. The same problem will also confront plastics largely because oil is both the energy source and the feedstock.

Other materials which are much more readily available will supplant many applications, a continuation of the competition for usage which has been intensifying over the past two hundred years. In such a steady state situation, i.e. no growth for many single materials, the availability and ease of recycling old scrap becomes vital and it is incumbent to design for ease of dismantling, identification and recycling.

Unfortunately, despite the relatively low additional energy required, the world's performance in recycling old or used scrap as distinct from new or reprocessed scrap is poor.

Dissipation

For many metals which are likely to be in short supply such as cadmium, cobalt, tungsten, the dissipation rate is very great, say 90%. Uses of metals which lead to such gross losses must be considerably reduced even if it means discontinuing their use as sacrificial protection for steels or because they are dispersed and lost in small pieces in the shapes of tool tips for machining or as drill bits and tool tips for mining operations throughout the world. But even for metals such as copper, the quantity of old scrap recycled is only about 20% of any current year's production. The best performance is for lead and antimony at about 30%. For plastics it is virtually nil and recycling of such old or used scrap back to its original properties is extremely unlikely without pyrolysis, which is bound to be expensive and of low yield.

Timing

The time lapse between the initial conception by a design engineer for a new product and the end of its useful life may be from five to thirty years. This period could easily run into the time when supplies of some virgin metals and materials will start to fall short, or be very expensive. This will come about because of reduced availability and a considerable increase in energy costs.

All the metals which are components of such products can, if coded and readily and safely identified and if they are easily dismantled, be recycled at low cost and for little additional energy. Such facilities would do much to enable recycling of old scrap to reach 60 to 70% of a metal heat and thus be one of the dominant sources of metals in most countries, assuming a static economy. Also, in twenty to thirty years' time, market forces should be more favourable for recycling of old scrap. For these underlying reasons, the engineer must start now designing products to encourage easy and efficient recycling of them as old scrap.

Summary

In the short term, major savings in energy can be made by improving processing efficiency, concentrating on those materials which can be recycled from old scrap without deterioration in properties and also those of low total energy content such as timber, concrete and steels. In the longer term, however, the total energy content per unit of property coupled with the cost per unit of property should be used to point the way to efficient utilisation of our overall resources, i.e. energy and materials.

In any effort to conserve energy, materials and therefore, costs, the following major factors must be emphasised:

1) The total energy data is only just being collected, collated and its significance appreciated. Despite the difficulties and vagaries it is fundamental and obviously of increasing cost significance. Such costing and energy estimates must be as near to finished shapes as possible to include the bulk of energy and labour in the yield and comprehensive added value.

2) Ultimate yield of the finished product can vitally affect total energy and cost.

3) The higher the value of a specific property the more efficient it usually is in energy terms but costs often increase more than proportionately.

4) The useful working life of most materials must be increased substantially except possibly those of low energy content such as concrete and timber.

5) Designers must now start designing most products for easy dismantling and identification of all components to facilitate efficient recycling.

There will be future shortfalls in certain key materials, and hence we should exploit every material to its optimum in total energy terms. In the light of such an overall review of the tonnage metals and materials of the world, and the development of specialist metals, materials and composites, Man could probably manage with fewer of the resources which are likely to be in short supply or demand excessive total energy. The reasons are that there is an abundance of some materials and many properties which are desirable overlap between them. Furthermore, the ingenuity of engineers and scientists ensures that there is always more than one way to achieve a desirable end whether it is to produce a structure or an instrument, even if at a relatively high cost.

With foresight and adequate facts available to the world community, and providing we conserve all our energy and materials' reserves sensibly, we need not be too downcast by the forecasts of the prophets of doom.

References

[1] K. S. B. Rose, Energy Audit Series No. 1, Iron Castings Industry. Department of Energy and Department of Industry, 1977.

[2] Energy Audit Series No. 6, Aluminium Industry. Department of Energy and Department of Industry, 1979, and B.N.F. Metals Technology Centre, private communication.

[3] H. Kindler and A. Nikles, Kunststoffe 70 1980, 12, 802 - 807.

[4] J. C. Hewgill, Watt Committee on Energy, Consultative Council, May, 1979, 6, "Evaluation of Energy Use, Now and Tomorrow".

[5] S. S. W. Lam, Thesis: "A Feasibility Study on Energy Accounting". University of Stirling, 1977, 25 and 41.

[6] R. McDonald, private communication.

[7] Reference (5) p.27.

[8] Noranda Metals Industries Ltd., Montreal, Canada, private communication.

[9] I. Boustead and G. F. Hancock, "Handbook of Industrial Energy Analysis", Wiley, 1979.

[10] International Materials Congress, Reston, Virginia, U.S.A., 26th to 29th March, 1979, "Materials Aspects of World Energy Needs".

[11] International Federation of Institutions of Advanced Study, Sweden, 1974.

[12] 9th International T.N.O. Conference, Rotterdam, Netherlands, 25th to 27th February, 1976, "The Energy Accounting of Materials, Products, Processes and Services".

[13] W. O. Alexander and P. M. Appoo, Design Engineering, 1977, p.59.

[14] H. H. Kellogg, Institution of Mining and Metallurgy, 18.4.1977, "Conservation and Metallurgical Process Design".

[15] Report of a NATO Science Committee Study Group, "The Rational Use of Potentially Scarce Metals", p.16.

Thermal Insulation of Buildings

By I. Dunstan and L. H. Everett
BUILDING RESEARCH ESTABLISHMENT, BUCKNALLS LANE, GARSTON, WATFORD, HERTFORDSHIRE WD2 7JR, U.K.

Introduction

In recent years there has been a growing awareness of the importance of thermal insulation of buildings. Not only does improved thermal performance of the elements of the structure - roof, walls, floors, fenestration - lead to savings in the cost of heating it also provides more even temperatures within the structure, thus higher levels of comfort to the occupants, and reduces the incidence of condensation in dwellings. The latter has been recognised in the drafting of Building Regulations, Part F, and amended in respect of non-domestic buildings by Part FF which came into effect on 1st June 1979.

This paper is concerned primarily with the means by which these standards can be achieved, the demands made on thermal insulation materials, the possible drawbacks associated with improved insulation - condensation, likelihood of rain penetration, additional fire risk - durability of the materials, and the need, by the introduction of new and proposed legislation, for a closer study of thermal insulation as a whole.

With a housing stock of some 19 million households in the UK, approximately 9 millions of which have cavity walls, the energy demand in the domestic field is substantial and the paper will be concerned largely with dwellings although the principles developed will be equally applicable to the industrial sector (eg factory premises, storerooms, etc), to educational establishments - schools, colleges, - to shops and offices, hospitals,and also in some respects to the transport industry for structures such as railway stations, bus garages, waiting rooms, etc.

In this paper the techniques for preventing heat loss are outlined, but not the manner in which these energy savings are used, for this varies widely with the prerequisites of the owner or occupier. For example, a survey of energy usage by local authority tenants in Scotland showed that on average only about half of the potential savings from improved thermal insulation were in fact realised as actual energy savings, the other half being used to produce higher temperatures[1]. It is also realised that occupiers providing little heat input into the building will not save much energy when the insulation is improved. Such socio-economic considerations are outside the scope of this paper as is also any consideration of the costs or cost-effectiveness of insulation methods for costs are highly variable even for the same system of insulation; in housing for example, it is related to the number of installations done at any particular site and to the timing of installation - during construction or after completion of the building. Similarly, any derivation of cost effectiveness requires an assumption of the cost of primary energy now and in future years, the probable rate of inflation, return on investment capital, and a realistic discount factor with which the present value of the insulation costs can be calculated.

Energy Usage

Official statistics for 1979[2] give data on the energy delivered to the UK economy and enable the gross energy consumption of the final users of energy to be deduced. From these statistics the primary energy consumption of the UK was 9.3×10^9 GJ and the energy delivered to the final user was 6.5×10^9 GJ, the difference representing losses in conversion and distribution in the energy supply industry. Figure 1 illustrates the four sources providing this energy and the consumption by various sectors of the economy. As will be seen, almost half the energy came from oil, a situation which is very different from that in the previous decade when 70% of UK energy came from coal. Over that decade, total energy consumption rose by an average of 1.9% per annum[3].

Taking account of the ratio of gross energy input to net energy delivered to users for various fuels, it is possible to derive from Fig.1 the gross energy consumed by final users by distributing losses amongst the final users in proportion to their use of various fuels. By this means, the energy used in the domestic sector is 29% of the gross input, while in the 'other users' sector (commerce, central and local government, agriculture, education etc) the figure approaches 14%. Since nearly all energy used by the domestic sector goes to provide heat and light within the building, which is also true

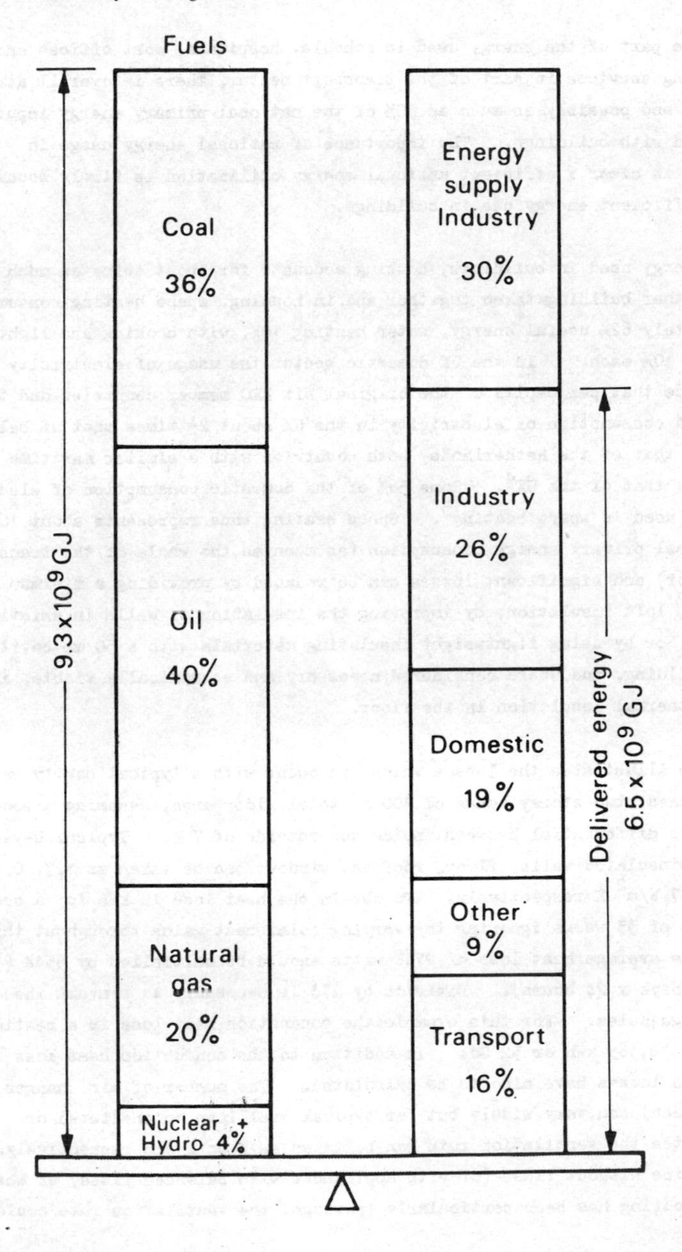

Figure 1 Energy balance for UK 1979 (including consumption by energy supply industry)

of a large part of the energy used in schools, hospitals, most offices and in building services in part of the transport sector, there is overall at least 40% and possibly as much as 50% of the national primary energy input associated with buildings. The importance of rational energy usage in buildings is clear : efficient national energy utilisation is firmly bound up with efficient energy use in buildings.

Of the energy used in buildings, housing accounts for about twice as much as all the other buildings taken together and, in housing, space heating consumes approximately 62% useful energy, water heating 18%, with cooking and lighting consuming 10% each. In the UK domestic sector the usage of electricity is about twice that per capita of the original six EEC member countries and the per capita consumption of electricity in the UK about 2½ times that of Belgium and twice that of the Netherlands, both countries with a similar maritime climate to that of the UK[4]. Some 36% of the domestic consumption of electricity is used in space heating[3]. Space heating thus represents about 16% of the national primary energy consumption (as much as the whole of the transport sector) and significant losses can be reduced by providing a minimum standard of loft insulation, by improving the insulation of walls in existing buildings, or by using lightweight insulating materials with a 50 mm cavity in new building, and, where considered necessary and economically viable, installing thermal insulation in the floor.

An example illustrates the losses which can occur with a typical cavity-wall, semi-detached, two-storey house of 100 m^2 total floor area, assuming a mean temperature differential between inside and outside of 7 K. Typical U-values for the uninsulated walls, floor, roof and windows can be taken as 1.7, 0.76, 2.1 and 5.7 W/m^2 K respectively. To obtain the heat loss in kWh for a heating season of 33 weeks ignoring the varying solar heat gains throughout this period, the average heat loss of 2748 watts should be multiplied by 5544 (33 weeks x 7 days x 24 hours). Division by 278 is necessary to convert the loss to gigajoules. For this example the conduction heat loss in a heating season is 14,365 kWh or 52 GJ. In addition to the conduction heat loss ventilation losses have also to be calculated. The number of air changes per hour (ach) can vary widely but for typical dwellings on sheltered or exposed sites the ventilation rate can be taken as 1 or 2 ach respectively. In properties without flues (or with appliances with balanced flues) or where draught-proofing has been particularly thorough, the ventilation rate could be ½ ach or less. To calculate the ventilation heat loss, the volume of the dwelling within the external structure (ie walls and ceiling) should be estimated, a deduction of the dead space - cupboards, partitions etc - say 10% made and this figure multiplied by the mean temperature difference, the hours in the heating season, the number of air changes per hour and the number of kWh required to raise the temperature of 1 m^3 of air through 1 K (0.33kWh). Assuming a ceiling height of 2½ m and a ventilation rate of 1 ach, for the typical dwelling considered previously the ventilation heat loss is calculated in Table 1 as about 10 GJ over the heating season. Hence, the total heat loss will be 62 GJ. The heat gain from the sun, from cooking and from the use of domestic hot water has to be subtracted from the calculated loss. This would be about 15 GJ per heating season, so the overall heat loss becomes 47 GJ for the uninsulated house.

Table 1 shows that in uninsulated houses on average the greatest heat loss occurs through the wall; only 23% is lost through the roof, 13% through single glazed windows, about 9% through solid floors in contact with the ground, and 17% by ventilation compared with about 38% through the walls. Insulating the walls therefore, provides one way in which the loss can be minimised.

For cavity walls this may be achieved by injecting or inserting insulating material into the air space. A wide range of materials are suitable for this purpose as will be considered later. An example of the effectiveness of this process may be found in the report of a 5 year experiment[5] on a detached two-storey house with conventional cavity walls and gas central heating. Measurements of the internal temperatures, ventilation rate, fuel consumption and external temperatures were taken some 2 years before the cavities were insulated with urea-formaldehyde foam, and for three years thereafter. For this property of about 100 m^2 floor area, the external walls having an outer leaf of facing bricks and an inner leaf of clinker concrete blocks, rendered and plastered, and a total window area of 21 m^2, the annual fuel saving in an average year was about 23 GJ amounting to a saving in fuel costs of 25 to 30%. Additional benefits were also obtained, in particular a more constant mean internal temperature and, even with the savings in fuel, a slightly higher mean internal house temperature (about 1K) for the heating season after the walls had been insulated.

TABLE 1(a) **Conduction and Ventilation Heat Losses for a Typical Uninsulated Semi-detached House**

1. Conduction Losses

Construction Element	Area (m^2)		U-value (W/m^2K)		Mean temp. difference (K)		Average rate of heat loss (W)
External walls	100	x	1.7	x	7	=	1190
Ground floor	50	x	0.76	x	7	=	266
Roof (including ceiling)	50	x	2.1	x	7	=	735
Windows	10	x	5.7	x	7	=	400
							2591

Therefore fabric heat loss over heating season = 2591 x 5544 = 14,365 kWh

52 GJ.

2. Ventilation Losses

For 1 air change per hour

Volume (m^3)		Mean temp. difference (K)		Length of heating season (hrs)		
225	x	7	x	5544	x	0.33

= 2881 kWh 10 GJ

where 0.33 is the kWh needed to raise the temperature of 1 m^3 of air through 1 K.

TABLE 1(b) **Possible Savings from Improved Insulation**

Construction Element	Action Taken	U-value (W/m^2 K)	Savings over Heating Season (GJ)
Roof	50 mm loft insulation	0.55	9
Walls	cavity insulation	0.50	17
Windows	double glazing	2.8	7
Floor	10 mm carpet and underlay	0.55	0.4

Methods of Insulation

For existing properties with cavity walls the most common method used to improve the thermal insulation is injection of organic or inorganic material into the cavity space. Urea-formaldehyde foam has been the cheapest and most widely used cavity insulation for many years. It is a low density cellular plastics foam which is produced by foaming together in the injection nozzle a mixture of water-based resin solution, a hardener-surfactant solution and compressed air. The foam, which has a consistency similar to shaving cream, is injected into the cavity space where it subsequently hardens and dries. As it dries it will normally shrink and this can lead to fissuring. The materials which arrive on site are a solution of the resin and the hardener, both of which occupy relatively small volumes so that the transportation requirements can easily be met by a small vehicle. The vehicle is equipped with a small compressor which supplies the air needed for foaming and provides the power for the drilling tools needed to form the pattern of holes in the outer leaf, through which the foam is injected. The wall cavities of buildings under construction may be filled by injection through holes drilled in the internal leaf of the wall before this is plastered so the time and cost of the whole process can be significantly reduced. Before filling the cavities, samples of the foam are taken to ensure compliance with BS 5617 and the drilling pattern is carried out in accordance with the requirements of BS 5618.

Polyurethane foam also provides equivalent thermal performance and involves the injection of a mixture of two liquid components into the cavity where they foam spontaneously and rise to fill the cavity space. The polyurethane foam adheres strongly to masonry and its shrinkage is less than for urea-formaldehyde foam but the cost is higher. As well as providing additional thermal insulation high density polyurethane foam has been used to stabilise cavity walls where the metal wall ties have corroded but for this purpose the materials cost is significantly higher.

Both granular and fibrous fills can also be used for new or existing buildings. The most well known granular fill is expanded polystyrene beads. These are white spheres, about 2 to 7 mm in diameter, and have extremely free running properties so few filling holes are required. The wall is usually drilled, with holes up to 65 mm diameter, at high level and under obstructions such as windows, and the fill poured or blown into the cavity. The free running nature of this insulant can lead to an unnoticed escape of material through holes in the inner leaf, for example around the ends of timber joists or service pipes, ducts, bathwastes etc, so particular attention must be given to seal these locations. Polystyrene granules, made by shredding waste eps bead board, are about the same size as beads but because of their angularity they are less free flowing so there is less risk of escape from the cavity. The relatively recent introduction of bonded beads, where the spherical beads are thinly coated with adhesive as the fill enters the wall, also restricts the escape of the fill. Polyurethane granules can also be employed as an effective fill material. The granules are irregularly shaped, usually between 5 and 20 mm across and are produced from waste rigid polyurethane foam. An inorganic fill, perlite, is available for cavity fill. For this purpose it is normally supplied with a particle size less than 1 mm and coated with silicone to prevent moisture absorption. The siliconised perlite particles are blown into the cavity but special equipment is needed to avoid fracturing the particles and creating a high percentage of dust.

Unlike the foam products, the particulate fills are all dry materials so problems of increasing the moisture content of walls and of extending the drying-out period of new construction do not arise. The rock or glass fibre fills also fall into this category. Rock fibres used for cavity wall insulation carry a coating of water repellent and are blown as bundles of fibres (tufts) into the cavity region where they form a water repellent mat. Although the second most widely used fill in this country, the number of installations is much lower than for UF foam. The raw material cost is also higher and bulk transport is required to bring to site the whole volume of fill; hence rock fibre fills are more expensive than UF foam. Glass fibre has been a relatively recent introduction which is installed in a similar manner to rock fibre. The glass product is less dense than rock fibre when installed so materials costs are lower.

All the above materials may be injected into cavity walls during construction but other methods can be considered at the design stage. This has the additional advantage of ensuring that detailing and materials are compatible with the insulation material and have regard to the exposure of the building. For new walls, the range of possible insulants can be extended to include glass and rock fibre slabs and expanded polystyrene boards. The glass and rock fibre slabs have excellent insulation properties but need to be installed with care to prevent the possible penetration by water at joints. Wide gaps between adjacent slabs must be avoided and any mortar droppings from the brickwork prevented or removed. Good supervision is essential and the outer leaf built up first to a height of at least one slab and the inner leaf built against the insulation mat. This order of construction prevents extrusion of the mortar in the outer leaf into the joints between adjacent slabs. Expanded polystyrene insulation board is normally fixed to the inner leaf in the cavity and is generally about 25 mm thick so that a clear cavity is retained behind the external wall. Special fixings and wall ties are available to prevent disturbance of the sheet so that they do not lean across the cavity. It is difficult to keep a narrow cavity free of mortar droppings so recommendations have been made to leave a 50 mm clear cavity with the boards in place. Agrement certificates for these sheets specify acceptability for different exposure conditions, depending on the width of the remaining cavity.

Timber frame constructions generally carry the insulation on the inside of the sheathing in the form of a glass fibre quilt fixed between the timber framing. In this position the insulation is removed from any moisture penetrating the external wall (often facing brickwork with a conventional 50 mm cavity), but under extreme humidity conditions may become moist from condensation. Other methods of reducing the heat loss from these structures may be used in addition to the quilt, eg reflective barriers or aluminium-faced plasterboard linings.

External Insulation

A large proportion of the UK housing stock is of solid wall construction and various techniques for applying external insulation have been developed. In almost all cases the external appearance of the structure will be changed and how these changes are brought about depends on the specific requirements of the owner and his architect. The choices range from use of a proprietary system to one of individual design. Existing cladding systems such as tiles, weather-boards or detachable sheet material can be upgraded by incorporating additional insulation material in sheet form behind the cladding system with minimum effect on appearance and would be the option selected in some, albeit few, situations. Proprietary systems, installed by experienced personnel, are likely to be used in the majority of situations and the opportunity taken of replacing major items such as badly weathered windows, sills or providing new window openings as required at the same time. Often, on older properties, the appearance can be markedly improved in the process.

There are, in general, four types of external insulation system. These can be classified as the direct or indirect mechanically fixed systems, insulated renderings and adhesive fixed systems. In the direct mechanically fixed system, composite boards of thermal insulation material bonded to a waterproof breather paper and wire lathing is fixed directly to the wall by special fixing pins. The wire lathing acts as a key and a reinforcement for conventional rendered finishes. Care needs to be taken to ensure that the metal lathing is not prone to corrosion and that the external rendering is sufficiently thick and sound so as to protect the metal fully. Insulation efficiency is also impaired if the fixings compress the insulant, so the choice of insulant and its density is important to achieve maximum effect. Insulation thicknesses of 25 or 50 mm are usually employed. The thermal conductivity of the fibrous (glass or mineral wool) insulants normally used is around 0.036 W/m K so that 50 mm of insulant would give a 'U' value for a single brick wall of about 0.5 W/m^2 K. The slabs of insulant are layered so that any water entering the material flows downwards and does not tend to cross the insulation. Round headed nylon dowels at 600 mm centres horizontally and 300 mm vertically are used to fix the insulant and mesh which is then rendered with a normal cement:lime:sand mix. This is the most widely used system in the UK and has been under continuous development since 1975.

The indirect mechanically fixed system is a development of the traditional cladding of clay tiles or weatherboard fitted on battens. The thermal insulation is added between the battens. Alternative claddings are rendered metal lathing or panels of glass reinforced cement. Expanded polystyrene board, either as beadboard or as extruded board up to a thickness of 110 mm is available.

Insulated renderings use lightweight aggregates, such as perlite or expanded polystyrene beads, in a cement matrix. The thermal conductivity of the insulated render is of the order of 0.08 W/m K and it is usually applied to a thickness of 60 mm. The density is about 300 to 400 kg/m^3. In practice, a thick layer of render is built up on the wall by trowelling or spraying and a final decorative and protective finish is applied. If the risk of impact damage is slight the protective finish is a dense, textured paint coating but where the risk is greater the rendering has to be reinforced. The advantage of the system is that it is very flexible in that it can be applied to curved walls and can be modified to overcome detailing problems, but it has the disadvantage that the substrate has to be carefully prepared (as for traditional rendering) and provides less insulation value than a similar thickness of other insulants. The overall performance is also dependent on the durability and maintenance of the final external finish which serves to protect the render from rain.

In the adhesive fixed system, boards of lightweight insulant are stuck on to walls with a special adhesive. A reinforced coating is applied over the boards to improve impact resistance, provide weatherproofing and inhibit cracking from thermal or moisture movements. The insulating board used is often extruded or bead polystyrene or expanded pvc. Often mechanical and adhesive fixing to the wall is employed. In one system the attached boards are rendered with a mixture of specially formulated adhesive containing white Portland cement into which a sheet of tough, plastics coated woven glass fabric is embedded. After hardening it is decorated with a synthetic plaster applied by trowel or spray. This system was patented in 1959 and has been applied to buildings in Germany since 1963. More recent systems of this type dispense with the glass fabric and use 6 mm of GRC (glass reinforced cement) rendering instead; which gives good impact resistance. A decorative finish is applied above the GRC.

A further, relatively inexpensive modification of adhesive bonded systems is where polyurethane foam is sprayed directly onto the substrate. The foam is self adhesive to roughened surfaces but will deteriorate on weathering unless protected by paint or other coatings to protect it from the action of UV light and absorbed water. It can be used as additional insulation behind cladding but its main application has been on large roof areas on factories and similar industrial buildings.

Internal Insulation

Internal insulation may be considered desirable where the external facade has to be preserved for historic or aesthetic reasons. Possible internal treatments include the following, (the bracketed figures giving the improved thermal transmittance obtained over an existing 220 mm solid brick wall with 16 mm plaster internal finish where the 'U' value is 2.13 W/m^2 K)

a) wallpaper on aluminium backed plasterboard on battens (1.14)
b) wallpaper on plasterboard on battens with 25 mm insulant between battens (0.90)
c) as b) above but with 50 mm insulant (0.61)
d) wallpaper on composite plasterboard and 13 mm foamed insulant adhesively fixed to wall (1.04)
e) as d) above but with 19 mm sheet foamed insulant (0.85)
f) plaster on 25 mm foamed polystyrene board adhesively fixed to wall (0.86)

Treatments applied to the inside of the external walls are generally less expensive than those applied to the outside but the thermal benefit that might result from the position of the insulation and loss of occupancy for internal treatments have to be examined as below:

Advantages and Limitations of External and Internal Treatments. As indicated above it is not possible or desirable to lay down clear advantages which apply to every specific location. In determining particular schemes however the following points need be assessed:-

a. Thermal insulation. The thermal transmittance of an uninsulated

220 mm thick solid wall would be about 2.13 W/m^2 K. This can be reduced to around 1.0 W/m^2 K by adding 50 mm of lightweight render externally or 20 mm insulant behind plasterboard internally. Further reduction to 0.5 W/m^2 K is possible by increasing the insulant thickness to 50 mm.

b. Protection and appearance of the wall. External insulation, correctly applied, reduces the moisture content of the structure by protecting the wall from driving rain. Internal insulation results in no improvement in this respect. Against this has to be balanced the risks of interstitial condensation, within the insulation or on cold surfaces with external insulation and condensation within the now cooler wall with internal treatments. Clearly a vapour barrier will overcome this problem but it has to be correctly positioned and efficient to achieve the optimum effect not only in regard to interstitial condensation but also as to where the water vapour will eventually dissipate if the wall does not continue to give the 'sink' capacity it has hitherto provided. There is scope for further examination of this subject.

The appearance of the wall will have been changed by external treatment. This may prove unacceptable but in many areas of the UK, in Scotland for example, there already exists a strong tradition of rendered buildings. Often, particularly in rehabilitation work, the final appearance of the dwelling following external insulation will have been improved.

c. Thermal capacity of the wall. External renderings preserve the thermal storage capacity of the wall resulting in more even temperature conditions inside the dwelling and so improving the comfort level of the occupants. Heating the walls consumes energy and this can be minimised by internal lining. Moreover, internal lining results in higher internal temperatures of occupied spaces being attained much more quickly for lower energy consumption, which is attractive for short term intermittant occupancy, but also allows the room to cool down more quickly when the source of heating is removed. Attention needs to be given therefore to user practice in considering different systems. The effects of thermal shock on the structure and the wide range of temperature variations experienced with internal treatments have also to be taken into account.

d. Cold bridging. Here it is far easier to solve problems of thermal discontinuity at floor/wall and wall/wall intersections by external insulation which masks cold bridges.

e. Condensation. This is reduced with external insulation and the well known problem of providing an effective vapour barrier as required for internal insulation is avoided.

f. Detailing. Externally, the need to prepare the wall, to deal with narrow projections on eaves and verges, with window openings, window and door reveals, also to reposition rain water pipes and to overcome difficulties associated with house extensions, conservatories etc raise many problems. They can be overcome but at additional cost. Internally the difficulties are not so acute.

g. Cracking. Thin external renderings will respond more quickly to changes in the weather, particularly to frosts and solar radiation, which can induce thermal stresses and eventual cracking. On large areas of wall some cracking is probably inevitable; what remains to be determined is the extent of cracking which is acceptable and methods of controlling crack widths. Reinforced renderings may provide one solution but for traditional renderings workmanship and adequate controls on the quality of materials are important.

h. Others. Other limitations on external treatments concern the likelihood of damage due to vandalism, the overall durability of protective paint treatments where applicable, the delays in completion due to bad weather and the possible fire hazard where combustible organic materials are used or where dense smoke or hazardous fumes are evolved in a fire situation. The latter applies even more so to organic-based insulants used internally which require thorough protection (usually by plasterboard) to give a sufficient degree of fire resistance. Advantages which external insulation systems possess are the lack of disturbance to the occupants of the building and the preservation of internal space.

Roof Insulation

In energy terms the loss from roofs is exceeded only by that lost through uninsulated walls hence improvement in thermal performance is to be encouraged. It is convenient to divide roofs into pitched and flat types since the insulation materials and the design to minimise condensation differ.

In pitched roofs the attic space provides access for installing insulation between the ceiling joists. The insulant most often used in housing is glass fibre quilt which is available in several widths and thicknesses to give the required insulation value and convenience in placing. This insulant has been in use for many years, together with rock fibre, in quilt form but more recently blown-in glass and rock fibres have been introduced. Over the last few years blown-in cellulose fibre insulation has appeared on the market and become commercially competitive with the mineral fibre products. Exfoliated vermiculite, perlite and expanded polystyrene beads have also gained acceptance as convenient alternatives to the traditional quilt material.

As regards thermal performance, the insulants give similar insulation values (k of about 0.032) except for exfoliated vermiculite which has a higher thermal conductivity of 0.062. The difference in thermal properties is reflected in the thicknesses of insulation required under the Homes Insulation Act 1978[6]. Installing insulation at loft floor level results in cooler conditions within the loft space and, with warm moist air rising from the living areas below, increases the relative humidity of the air and the risk of condensation on the underside of the roof. Continuous measurements of temperature and humidity of loft spaces have shown that over a normal heating season condensation does occur in roofs and for a considerable part of the heating period, a relative humidity above 80% is achieved in unventilated lofts. Ventilation serves to reduce the humidity and should condensation occur, to increase the rate of evaporation. Well insulated loft spaces should therefore also be well ventilated to avoid condensation drips onto the insulation and eventual damage to the ceiling or roof structure. Information is available[7] whereby the internal temperatures and dewpoints in roofs which do not have ventilated air spaces can be calculated so that the risk of condensation can be assessed.

The need to minimise the risk of condensation is nowhere more apparent than in the flat roof. Such moisture, especially if entrapped, can affect the properties of many materials used in roof construction, including the thermal insulants. Various options[8] are available to the designer of flat roofs to meet the basic requirements of the roof to protect the occupants and the structure from rain, wind and snow and from the extremes of temperature, noise etc of the external environment. One option is to place the insulation above the ceiling containing the vapour barrier but separated from the roof deck and its waterproof layer by a cold cavity. The advantages of this "cold deck design" are that the waterproof layer is fully supported by the roof deck, the waterproof layer is readily repaired or replaced, loose insulation material can be used above the ceiling and any moist vapour from occupied spaces can be ventilated from the cavity. The disadvantages of cold deck design lie in the difficulties of providing efficient ventilation of the cavity and the need for an efficient, durable vapour barrier at ceiling level, ie below the insulation.

A second option provides for the thermal insulation to be placed above the roof deck and protected from the external environment by a waterproof covering. In this "warm deck" design the roof deck has on one side the insulation, but separated from it by the vapour barrier, and on the other side an unventilated warm cavity which is immediately above the ceiling. The advantages of this design are that the roof cavities may be eliminated unless they are used to convey services, no ventilation is required, fire barriers are easily accommodated and the vapour barrier can be provided on the deck to prevent condensation. The disadvantages are that the insulation has to support maintenance traffic and equipment, it has also to be securely fixed to the deck to resist wind suction forces, replacement of the waterproof layer may involve sacrificing insulation, and failures in the waterproof layer are difficult to locate and repair.

A third option is the warm deck design, 'upside down', construction where the thermal insulation is positioned above the waterproof layer but below an external surfacing such as gravel. The roof deck supports the waterproof membrane, insulation and external surfacing. The advantages of the inverted roof are the protection provided to the waterproof membrane from sunlight and maintenance traffic and the elimination of a separate vapour barrier. The disadvantages are the difficulty in draining the roof, the need to secure the insulation from wind suction forces without impairing the efficiency of the waterproof membrane, the inaccessibility of this membrane for repair, and the need to choose a 'waterproof' insulant to maintain optimum thermal characteristics in a wet environment[9].

All of these roofing systems can perform well if the materials are carefully chosen. Serious problems were encountered with some of the earlier insulating materials which had excessive thermal movement causing cracking and perforation of the roof coverings. Dimensional stability of lightweight construction, particularly the movement of the thermal insulant, has to be accommodated in the design of the system as a whole so that the waterproofing layer remains intact.

The importance of good thermal design cannot be overemphasised — the guiding principle is to ensure that the vapour barrier is always on the warm side of the insulation layer. While this is not difficult to achieve in roofs constructed to the 'cold deck' design, poor vapour barriers and lack of cavity ventilation introduce the risk of continued presence of condensation over many months of the heating season with consequential durability problems.

Floors

Although small, the heat losses through solid floors in buildings can be effectively reduced at low cost by providing insulation during construction. The

heat flow from inside the building to the outside air is shown diagramatically in Figure 2. The greater the distance heat has to travel the smaller the quantity of heat lost. Providing a layer of insulation above the damp proof membrane reduces the loss to very low levels but is expensive and only likely to be used when electric underfloor heating is employed. A less costly alternative is the positioning of a horizontal strip of insulant at least 1 m wide around the perimeter of the floor between the ground floor slab and hardcore in conjunction with a vertical strip through the full thickness of the slab. Simplest of all is a vertical layer of insulant extending from finished floor level down to a depth of not less than 250 mm but with advantage down to the top of the foundations. Rigid sheet insulation materials are often used since these are more impermeable to water penetration into the insulant with consequent reduction of insulation value.

For suspended ground floors additional insulation is often provided in the form of a continuous layer of semi-rigid or flexible insulant laid over the timber joists or semi-rigid material between the joists[10].

Basic Properties of Insulants

The thermal properties of cellular plastic insulants are given in Table 2. As can be seen, the thermal conductivity is between 0.030 to 0.040 W/m K for most of the commonly used materials; the coefficient of thermal expansion is fairly high, the compressive strength quite low and the water vapour resistivity of the board materials such as extruded polystyrene, expanded pvc and rigid polyurethane sufficiently high to consider them in the bulk form, if not as vapour barriers, as low water absorbers.

Table 3 lists the thermal properties of some flooring and roofing materials while Table 4 gives some applications for various insulants and comments on one very important property - that of fire performance.

Materials may be classified as combustible or non-combustible. Combustible materials in the form of fibres or foams will ignite more easily than the same material in board form but board materials used as internal linings may contribute to the growth of a fire by spread of flame across their surface. Organic insulants will produce toxic combustion products often accompanied by dense smoke. The combustibility of thermal insulation materials may constitute a hazard but it is the form and positioning of the insulant that affect the risk. Plastics materials fall into two main groups, the thermoplastics, which soften

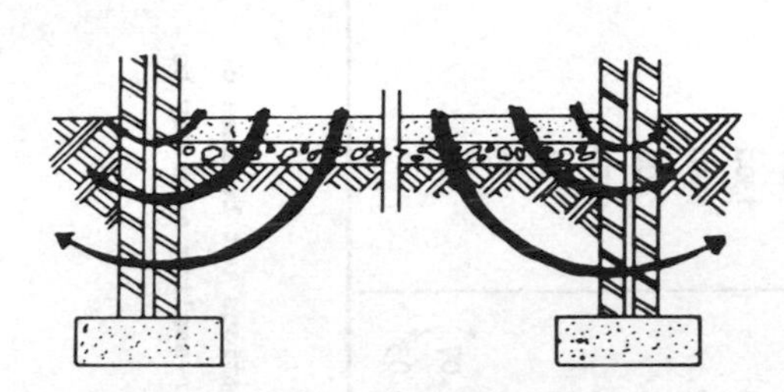
Fig. 2a Heat flow through ground slab

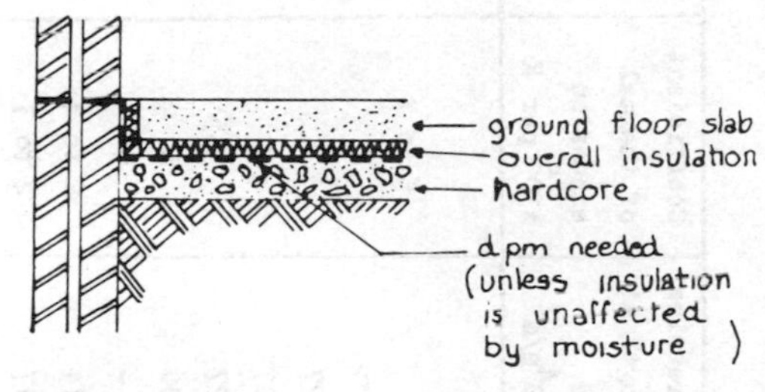

Fig. 2b Overall floor insulation

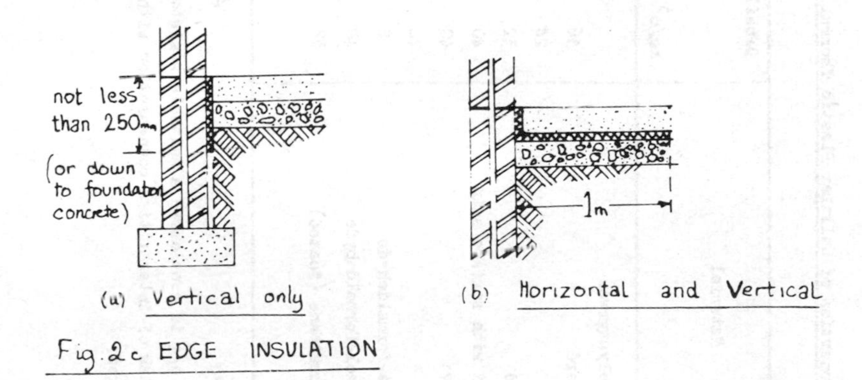

(a) Vertical only (b) Horizontal and Vertical

Fig. 2c EDGE INSULATION

Figure 2 Thermal insulation of ground floors

TABLE 2 : Properties of Cellular Plastic Thermal Insulation Materials

Material	Density kg/m^3	Compressive strength at yield $x\ 10^4 N/m^2$	Coefficient of thermal expansion $x10^5$ per K	Thermal conductivity W/m K	Water Vapour resistivity(1) MNs/g
Expanded polystyrene					
- bead board	16	7	5 to 7	0.035	270
	24	12	7	0.033	420
- extruded	32	27	7	0.035	1300
- extruded with surface skin	40	27	7	0.032	1300
Expanded pvc	40	27	3.5	0.035	800
	72	90	5	0.043	1300
Foamed urea-formaldehyde	8	-	9	0.038	22
Foamed phenol-formaldehyde	48	14	2 to 4	0.036	35
Rigid polyurethane (foamed)	32	17	2 to 7	0.03 - 0.07* 0.01 - 0.03†	360

* CO_2 blown   † Fluorocarbon blown

(1) 15 MN s/g is considered a suitable resistance for a vapour barrier in building applications but because of the risk of interstitial condensation within a cellular material, an additional vapour-sealing skin may be required.

TABLE 3 : Thermal Properties of some Flooring and Roofing Materials

Material	Density kg/m^3	Conductivity (k) W/m K	Resistivity (1/k) m K/W
Carpet		0.055	18.2
- wool felt underlay	160	0.045	22.2
- cellular rubber underlay	400	0.10	10.0
Cork flooring	540	0.085	11.8
Fibreboard	300	0.057	17.5
Glasswool quilt	25	0.04	25.0
Linoleum (BS 810)	1150	0.22	4.6
Mineral wool			
- felted	50	0.039	25.6
	80	0.038	26.3
- semi-rigid felted mat	130	0.036	27.8
- loose, felted slab or mat	180	0.042	23.8
Cellular plastics			
- phenolic foam	30	0.038	26.3
	50	0.036	27.8
- polystyrene, expanded	15	0.037	27.0
	25	0.034	29.4
- polyurethane foam	30	0.026	38.5
Pvc or rubber flooring		0.40	2.5
straw slab, compressed	350	0.11	9.1
Wood			
- hardwood		0.15	6.7
- plywood		0.14	7.1
- softwood		0.13	7.7
Wood chipboard	800	0.15	6.7
Woodwool slab	500	0.10	10.0
	600	0.11	9.1

Source : IHVE Guide, Book A : 1970

TABLE 4 : Thermal Insulation Applications

Material	Applications	Comments	Fire Performance
Expanded polystyrene	Walls and skin roofs, thermal insulation of flat roofs and concrete floors. Used as wall and ceiling tiles, insulation of cold water pipework. Cavity wall and roof insulation	Good thermal insulation, low softening point, attacked by organic solvents. Can remove plasticiser from pvc insulation of domestic wiring	Burns fairly rapidly, softens and collapses. Flame retardant grades available
Expanded pvc	Lining of walls and skin roofs; semi-structural sandwich panels	Relatively strong; low water-vapour transmission; low softening point	Collapses, but burns with difficulty
Foamed urea-formaldehyde	Filling of cavity walls	Low physical and mechanical properties. High shrinkage	Resistant to ignition
Foamed phenol-formaldehyde	Lining of walls and flat roofs and as core material for sandwich panels	Good fire and high temperature characteristics. Relatively poor physical and mechanical properties	Highly resistant to ignition

TABLE 4 : Thermal Insulation Applications (Continued)

Material	Applications	Comments	Fire Performance
Foamed polyurethanes			
(i) Flexible	Shaped insulation for pipework, acoustic applications, sealing and jointing	Relatively high thermal insulation and water vapour transmission	Generally burns rapidly, producing thick, dark smoke
(ii) Rigid	*In situ* insulation of roofs; as insulating boards for use in roofs, walls, under-floor applications. Core material for sandwich panels. Insulating and stabilising cavity walls. Spray-up foam has been used externally with a weather-resistant coating for walls and roofs.	Excellent thermal insulation; fairly good high temperature and water vapour transmitting characteristics	Flame retardant grades available

and flow on heating, often before ignition occurs, such as polystyrene, polypropylene, pvc and acrylic polymers, and the thermosetting materials such as urea formaldehyde, polyurethane and polyisocyanurate, which do not soften but char on contact with the heat source.

The non-combustible thermal insulants are expanded perlite, expanded vermiculite, glass fibre, slagwool and rock fibre. The fibrous materials are bonded so as to form a mat or quilt which can be easily fixed or laid in position. Combustible materials include cellulose fibre, produced by shredding waste paper which is then milled with fire retardant chemicals to prevent smouldering, polystyrene which ignites on application of a large heat source producing molten droplets and heavy black smoke, wood fibre insulating board, and the remaining thermoplastic and thermosetting materials listed above[11].

In the roof space, loft insulation materials are not subject to any control as to their fire behaviour[12] and with combustible materials care has to be taken to avoid all possible sources of ignition such as overheated or defective chimneys, inset lighting fixtures, sparks from defective wiring, discarded cigarette ends or the misuse of other heat or light sources.

Combustible material in cavity walls is protected by the masonry walls in any fire situation. Urea-formaldehyde foam completely filling the cavity is unlikely to result in rapid or extensive flame spread if the wall is breached; with expanded polystyrene in new construction the restricted ventilation and the robustness of the walls restrict active flaming to the area near the heat source but smoke and decomposition products are likely to diffuse through the cavity or into loft spaces and habitable areas. Organic granular or bead products have similar performance.

Internal linings present a different situation and any exposed combustible and flammable insulant increases the fire hazard. The good thermal properties of the insulant encourage rapid temperature rise and accelerate flame spread, particularly with thermosetting resins. With thermoplastic materials, eg polystyrene ceiling tiles, the fire may result in the release of burning droplets which in turn ignite the contents of the room below. If polystyrene tiles are attached by overall adhesive to a non-combustible ceiling and left undecorated, tests have shown there is no serious fire hazard. If these tiles have to be painted only matt finish flame-retardant or water-based emulsion paints should be used.

Performance of Insulants

Insulants used in floors and lofts generally give adequate performance if the correct material is chosen, in respect of low water absorption (for floors in particular) and satisfactory fire properties. Organic materials in lofts can cause concern on two issues. One is in regard to the reaction between polystyrene and plasticised pvc (as used for electrical cable insulation) where the polystyrene causes migration of plasticiser to the surface of the pvc. Although this does not increase the fire risk, it causes loss in flexibility of the pvc. For electric cables, disturbing the lay of the wiring may cause the embrittled insulation to crack. The reaction also results in shrinkage of the insulation and the possibility of exposing the ends of wires terminating in ceiling fittings. Direct contact between electricity cables and polystyrene loft insulants should be avoided by interposing a suitable barrier between the materials or relocating the wiring.

The other concern is the risk of corrosion of metals from the chemicals used to restrict smouldering of cellulose fibre loft insulants. When first introduced into this country the chemicals incorporated into the cellulose fibres consisted of a dry powder mix of aluminium sulphate, boric acid and borax. Cellulose fibre so treated was corrosive to copper, zinc and galvanised steel. Tests on the material showed that zinc losses were high under normal loft conditions so that the galvanised layer on steel components in the loft could be breached in a few years. These tests are now being developed, taking into account the inhibitive action of some of the chemicals now employed (eg borax) and the sensitivity of building components such as galvanised steel trussed rafter connector plates to corrosion. Such testing has to be short term and the results obtained capable of extrapolation for a possible life of 60 plus years.

With cavity wall insulation the performance of the insulant in resisting rain penetration has to be considered. The cavity wall was originally conceived to prevent dampness reaching the inside surface but all injected fills bridge the cavity and therefore provide a potential route by which rainwater can cross from the outer to the inner leaf. Different types of fill vary in this respect; the particulate fills (eg blown-in fibrous material) are made water repellent in processing, the barrier fills (eg foam) have very low permeability to liquid water so that rain penetration should not occur by capillary action through the pores. Water can pass quite freely however through any larger voids or cracks which occur in the fill.

Many thousands of dwellings have been insulated without any adverse affect on rain penetration of the walls but some instances of penetrating dampness have occurred. To quantify the risk a survey of some 32,000 Local Authority dwellings using UF foam fills has been conducted by BRE. The survey showed that the overall risk is of the order of 0.2% for dwellings filled several years after construction, with the greatest incidence of problems in those regions of the country where exposure conditions can be defined as severe. Filling with UF foam during construction incurs a higher risk of about 3.3%.

Experimental testing is essential to assess the merits of the various fills available. Such tests are usually conducted on specially prepared wall panels but these have several serious disadvantages over full scale tests. Panels undergoing such testing usually are less than one storey high whereas most housing is of two-storey construction. They are also essentially simple structures with none of the complexities of real buildings which affect completeness of filling and subsequent shrinkage or settling characteristics of the insulant. Panel tests can demonstrate however that single-leaf masonry is seldom leaktight so that under storm conditions water will flow continuously down the inside of the outer leaf. In assessing the performance of fill material, full scale tests under conditions which are controlled but also representative of normal weather conditions are required. Under these conditions the Driving Rain Index (DRI), which is the product of the annual average rainfall and annual average wind speed, and its relationship to the risk of rain penetration for filled and unfilled cavity walls, can be assessed. Such tests done at an estate in Essex on existing construction have enabled the quantity of water crossing the filled cavity to be correlated with the amount incident on the vertical surface. The results of these tests have been published[13] so a brief description only will be given in this paper.

In these tests water was applied to selected areas of the walls at a rate of 30 litres per hour (about 3.3 $l/m^2/h$) for 6 hours on each of four successive days for filled walls and until steady conditions were reached for unfilled walls. The test conditions were fairly severe but were aimed to provide a reasonable design condition, ie one which should not cause significant trouble.

Leakage rates of the outer leaf were of the order of 0.5 $l/m^2/h$ for a spray rate of about 3.0 $l/m^2/h$ and the volumes of water entering and crossing the cavity could also be determined for each fill. It is worthwhile noting here that

a) all the filled walls passed substantial volumes of water to the inner leaf

b) the 'best' filled wall passed about 4 times as much as the worst unfilled wall, and the worst filled wall about 28 times as much.

Both laboratory and site tests have established the specific moisture conditions which the fills have to withstand and provide the background from which performance requirements for successful fills can be drafted. The results obtained from this kind of testing haveassisted the preparation of a BSI Code of Practice on UF foam for cavity insulation and will be used in the Standard now under consideration on fibrous fills.

Concern has recently been expressed on the evolution of formaldehyde gas from UF foam. Experience in North America,where the specification and usage of the foam is not controlled as in the UK, has indicated that formaldehyde gas can be evolved for long periods after installation. The Consumer Product Safety Commission in the USA has held hearings following reports of adverse health effects. The Chemical Industry Institute of Toxicology reported that in animal tests where rats had been exposed to high concentrations of formaldehyde vapour, in excess of those likely to be found in housing in the UK and well above the Threshold Limit Value of 2 ppm, some of the rats had developed nasal cancer. As a result of the information brought before the Commission a ban on the use of UF foam for cavity insulation has been recommended. The product is already banned in Canada and in the State of Massachusettes.

North American practice differs from that in the UK in that cavities tend to be wider so more foam is needed to fill them, the construction is predominantly timber frame with an impermeable outer skin which does not permit the vapour to diffuse outside the building, and the ventilation rates tend to be lower than in homes in the UK.

Although there is considerable debate on the mechanism of formaldehyde gas release from the foam, three factors contribute to it:

1. Urea formaldehyde resin has a free formaldehyde content limited by the British Standard to less than 1% by weight of the liquid resin.
2. During the early stages of curing formaldehyde is produced, the majority occurring in the first few hours.
3. At high humidity and moderate temperatures, UF foam is hydrolysed and releases formaldehyde gas even after months of air curing.

While the UF foam used as cavity fill in the UK has the potential to release formaldehyde gas, the incidence of complaints is low, possibly because of the close control on resin formulations, the type of construction in this country - which restricts entry of the gas into living areas in the home, the care taken by installers in filling non-traditional walling,and a voluntary restriction by the industry to avoid using thick layers of UF foam in the loft space.

The Future

Considerable progress has already been made in reducing heat losses from buildings in this country. So what of the future? There is general acceptance of the need to encourage insulation of loft spaces and lagging of hot water pipes and storage vessels (EEC Heat Generator Directive) but there is scope for even greater application of existing thermal insulation measures, particularly for cavity and solid walls. The market potential is great with 19 million dwellings in this country, only about 1.5 million of which have benefitted from full thermal insulation.

Technically, the experience to date suggests where improvements are needed - in the provision of insulants which satisfy not only the thermal requirement but are also safe on health grounds, free from fire hazards, non-corrosive and resistant to the climatic conditions experienced throughout the British Isles.

Other areas for insulation not covered in this paper, such as metallised glazing to reduce losses through windows, effective controls within occupied spaces, the use of heat pumps, solar heating, and particularly the technology for storing heat from ambient sources for long periods, present a technological challenge which industry may find rewarding.

Above all there is the practicability of introducing these measures into the building structure which calls for the good-will of the owner, the ingenuity of the designer and the co-operation of the supplier in producing the required technology.

Addendum

This paper is the responsibility of the authors and does not necessarily reflect the views of the Establishment.

References

1. BRE Digest 190 'Heat losses from dwellings' June 1976
2. United Kingdom Energy Statistics, HMSO, 1979
3. BRE Current Paper 56/75 'Energy conservation : a study of energy consumption in buildings and possible means of saving energy in housing' BRE, 1975
4. BRE Digest 191, 'Energy consumption and conservation in buildings' July 1976
5. BRE Current Paper 20/74 'Cavity insulation of walls : a case study", 1974
6. Homes Insulation Act 1978, HMSO
7. BRE Digest 110 'Condensation', March 1972
8. BRE Digest 221 'Flat roof design; the technical options', January 1979
9. BRE Current Paper 1/75 'Avoidance of condensation in roofs', 1975
10. BRE Digest 145 'Heat losses through ground floors', September 1972
11. BRE Digest 233 'Fire hazard from insulating materials', January 1980
12. BRE Current Paper 2/80 'Possible fire hazards of loft insulating materials', 1980
13. Full scale testing of the resistance to water penetration of seven cavity fills, Building and Environment, 15, pp 109-118, 1980

British Standards

BS 874 : 1973	Methods for determining thermal insulating properties, with definitions of thermal insulating terms
BS 2972 : 1975	Methods of test for inorganic thermal insulating materials
BS 3533 : 1962	Glossary of terms relating to thermal insulation
BS 3536 : Part 2 : 1974	Asbestos insulating boards and asbestos wallboards
BS 3837 : 1977	Specification for expanded polystyrene boards

BS 3869 : 1965	Rigid expanded polyvinyl chloride for thermal insulation purposes and building applications
BS 3927 : 1965	Phenolic foam materials for thermal insulations and building applications
BS 3958	Thermal insulating materials
BS 3958 : Part 1 : 1970	85 percent magnesia preformed insulation
BS 3958 : Part 2 : 1970	Calcium silicate preformed insulation
BS 3958 : Part 3 : 1967	Metal mesh faced mineral wool mats and matresses
BS 3958 : Part 4 : 1968	Bonded preformed mineral wool pipe sections
BS 3958 : Part 6 : 1972	Finishing materials, hard setting composition, self setting cement and gypsum plaster
BS 4841	Rigid urethane foam for building applications
BS 4841 : Part 1 : 1975	Laminated board for general purposes
BS 4841 : Part 2 : 1975	Laminated board for use as a wall and ceiling insulation
BS 5422 : 1977	Specification for the use of thermal insulating materials
BS 5608 : 1978	Specification for preformed rigid urethane and isocyanurate foam for thermal insulation of pipework and equipment
BS 5615 : 1978	Specification for insulating jackets for domestic hot water storage cylinders
BS 5618 : 1978	Code of Practice for the thermal insulation of cavity walls (with masonry inner and outer leaves) by filling with urea-formaldehyde (UF) foam
BS 5803	Thermal insulation for pitched roof spaces in dwellings
BS 5803 : Part 1 : 1979	Specification for man-made mineral fibre insulation mats

Definition of Terms Used

Primary or gross energy : The calorific value of the raw fuel, oil, coal, natural gas, nuclear and hydro-electricity, which is input into the UK economy. Nuclear and hydro-power are used only for generation electricity and the normal convention is adopted of attributing the primary energy of these inputs as equivalent to an efficient coal-fired power station producing the same electrical output.

Delivered or net energy : The energy content of the fuel actually received by the final consumer.

Useful energy : The energy required to perform a given task. The ratio of useful energy to net energy represents the efficiency of the appliance in use.

Gigajoule (GJ) : The unit of energy which equals 10^9 joules and corresponds to about 278 kWh or 9.5 therms.

Thermal conductivity (k) : A measure of a material's ability to transmit heat, expressed as heat flow in watts per square metre of surface area for a temperature difference of 1 K per metre thickness.

Thermal resistivity (l/k) : A property of a material, independent of thickness. It is the reciprocal of conductivity and thus has units of metres degree centigrade per watt, m K/W.

Thermal resistance (R) : A derived function calculated as the product of thermal resistivity (l/k) and thickness in metres. It is expressed as m^2 K/W.

Thermal transmittance (U) : A measure of the ability of a constructional element to transmit heat under steady flow conditions. It is defined as the quantity of heat that will flow through unit area, in unit time, per unit difference of temperature between inside and outside environments. It is calculated as the reciprocal of the sum of the resistances of each layer of construction and the resistances of the inner and outer surfaces and of any air space or cavity. It is given in W/m^2 K . For a brick/block cavity wall the U-value is calculated from:

$$U = 1/(R_{int} + R_{ext} + R_{cavity} + R_{brick} + R_{block} + \ldots)$$

Reduction of Environmental Hazards Associated with Different Sources of Energy

By F. A. Robinson, C.B.E.
PAST PRESIDENT OF THE CHEMICAL SOCIETY AND OF THE ROYAL INSTITUTE OF CHEMISTRY, CHAIRMAN OF THE COUNCIL FOR ENVIRONMENTAL SCIENCE AND ENGINEERING

Now that everyone is aware that supplies of oil are limited and that the price will continue to rise, there is growing concern amongst ordinary members of the public about the sources of energy that will replace oil in due course. It is symptomatic of this concern that, on 4 March, 1981 The Times had three pages of authoritative articles on the subject under the heading of "Energy Futures" and the first of these articles reviewed what it called "benign energy sources". So the subject of my contribution to this Symposium is clearly one in which the public is now very interested, especially in view of the militant demonstrations frequently reported from the USA, France, West Germany and Japan against the extension of nuclear energy. So, what is to succeed oil as the main source of energy during the next two decades? Cost of course will be of paramount importance in determining which option will be chosen, but it is now clear that environmental considerations will also play an important part in the decision.

Coal

It is only 25 years since oil and natural gas overtook coal as the major source of energy in most parts of the world, and it was expected that by this time coal would have regained its former position, and that in this country for example, up to 170m. tonnes per annum would be produced by the year 2000. But several things have happened recently that will prevent this target from being reached; these include the importation of cheaper coal from abroad, the recession, measures taken to economise on the use of energy, and the action of the miners to prevent the closure of uneconomic mines. Of course, any developments that make more efficient use of energy, such as the use of waste heat from power stations for heating housing estates, are to be welcomed, though such developments may well cause further delay in the opening of the Vale of Belvoir mine in North East Leicestershire, regarded at one time as essential to the National Coal Board's plans. Undoubtedly however, the new mine shafts will eventually be sunk and the environmental damage to the countryside that was feared and anticipated will take place and there is no way of preventing it. And this will be repeated every time a new mine is opened.

Pollution by Coal Combustion

However the increasing use of coal as a source of energy will increase the damage to the environment in other ways, unless new methods of burning coal are adopted. In the past this country suffered severely from the black smoke emitted by the chimneys of the Industrial North and Midlands, and from the smog produced by the domestic fires of London and other cities. The Clean Air Act of 1956 and new technology eliminated pollution by black smoke, dust and grit, but pollution by other products of combustion remains, and coal is a more serious polluter than is oil. It produces larger amounts of sulphur dioxide for example, and at present in this country these are emitted from power stations into the atmosphere by means of very tall chimneys, along with oxides of nitrogen, hydrogen chloride and trace metals, where they are quickly diluted with air and widely dispersed. Unfortunately our prevailing winds send these gases across the North Sea to the Scandinavian countries where, say their inhabitants, they are washed out of the air as acid rain into their rivers and lakes, thereby poisoning vegetation and fish.

A recent OECD Report has indicated however that contamination by these gases is heaviest close to the centres of industrial activity, that is in the UK, Holland and Germany and rapidly diminishes with distance, and that in any event the acid gases of which the Scandinavian countries complain originate from a very wide area of Europe, including countries belonging to the Eastern bloc.

In the USA, where the coal produced contains more sulphur than does European coal, strict emission controls are in force requiring desulphurisation of flue gases, usually by scrubbing with lime or limestone. This is expensive, both in capital cost and operating cost, but in the USA the emission of sulphur dioxide will not increase greatly even if much more coal is burnt. Other constituents of flue gases from coal-burning that must not be overlooked are trace metals and metalloids such as arsenic, chromium, beryllium, cadmium, mercury, selenium and radionuclides; most of these are removed by the scrubbing processes adopted in the USA. As I have said, these processes are expensive and one is faced with the problem of how best to dispose of the slurry that is thereby produced.

<u>Fluidised bed combustion</u>

Interest in Europe is focussed on a different method of removing these gaseous products from coal during rather than after combustion. This is fluidised bed combustion, which is also being developed in the United States as it offers several advantages over gas-scrubbing processes. In principle, a fluidised bed is formed when a bed of finely divided particles is subjected to an upward stream of air of such a velocity that the particles become turbulently suspended and resemble a bubbling liquid. The bed is heated up by burners directed into the surface and, when a temperature of about 600°C is reached, crushed coal is introduced at the base of the bed and is burnt, continuously maintaining a temperature of 800-900°C, which is below the fusion temperature of most coal ash. A bag filter or electrostatic precipitator is used to collect the dust. When limestone or dolomite is introduced into the bed, sulphur dioxide is retained, the amounts depending on the type of limestone used, and if calcium oxide is used, over 90% of the sulphur is retained. Moreover at temperatures of 800-900°C, which are much lower than those in conventionally fired boilers, nitrogen in the air is no longer "fixed" and the amounts of oxides of nitrogen formed are governed solely by the nitrogen content of the coal. This lower combustion temperature also results in a reduction in the amounts of trace metals emitted. There are unlikely to be serious problems about disposing of fluidised bed ash, for it can be used as a concrete additive, in highway basecourse stabilisation and as a landfill material. Coal should no longer be regarded as a dirty fuel in comparison with oil or natural gas.

The "Greenhouse Effect"

The emission of large amounts of carbon dioxide however cannot be dealt with in the same way, although it could present a very serious environmental hazard of a quite different type. In 1974 the European Commission estimated that the energy requirements of the Community would be 2400m.t.c.e. in 1985, whilst the target set for coal production in the USA is 1500 m. tonnes per annum by the 1990's. So that, assuming that the other two countries with large reserves of coal, namely the USSR and China, will together produce some 2000m.t.c.e. by the 1990's, the amount of fossil fuel being burnt by that time will be well in excess of 6000m.t.c.e. per annum, since we have not taken into account the coal consumption from such other big producers such as Australia, South Africa or Eastern Europe, or the large amounts of oil still being consumed in many parts of the world. It is a fair assumption therefore that several thousand million tons of carbon dioxide will be emitted into the atmosphere each year from now on, and it is certain that only a fraction of this will be taken up by the oceans or used by the world's vegetation. In fact, the concentration of carbon dioxide has risen steadily from around 280 to 290 p.p.m. in the 19th century to around 330 p.p.m. in the early 1970's, and is currently rising at about $1/3$% per annum, largely owing to the burning of fossil fuels. A doubling of the concentration to about 600p.p.m. might well occur by the year 2050. Since carbon dioxide strongly absorbs infra-red radiation emitted by the earth's surface, higher concentrations of carbon dioxide would produce higher temperatures in the lower and middle atmosphere, by the so-called "greenhouse effect". In addition, increased evaporation of water from the earth's surface caused by these higher temperatures would increase the water vapour content of the atmosphere which in turn would absorb still more the terrestrial infra-red radiation, thereby reinforcing the heating due to the increased carbon dioxide.

Complex three-dimensional physico-mathematical models of the world climate have been developed in the Meteorological Office in this country and by three major groups in the USA, the only places in the world where this can be done, and although these cannot take into consideration all the variables involved, they can simulate the major features of the present world climate and enable the response of climate to various natural changes to be explored in a general way. At present it would appear that as a result of the probable increase in the carbon dioxide content to 600 p.p.m. during the next 50 years or so, the global annual average surface air temperature will increase by 2-3° with a maximum of 10° in polar regions in winter. Rainfall will increase by 5-7%, with considerable geographical and regional variations, increases being as much as 20% in the middle latitudes. However these beneficial increases in air temperature and in rainfall will not occur everywhere, and slight reductions in rainfall may be expected in summer in parts of Europe, Asia and North Africa. If these were to occur in a major food-producing region, the consequences could be serious. Much more research is needed before we can say with certainty what effect increasing amounts of carbon dioxide in the atmosphere are likely to have.

For the present therefore, there is no evidence that burning increased amounts of coal will have disastrous consequences, as is sometimes alleged, but if in 10 or 20 years time conclusive evidence is forthcoming that food production in some parts of the world may be seriously reduced, what then can be done to reverse the trend towards generating more and more carbon dioxide?

Nuclear Energy

The obvious answer is of course to use fuels that do not produce carbon dioxide, and the one about which we already have considerable "know-how" is nuclear energy. In spite of the strident and often militant opposition of anti-nuclear campaigners in many countries, a limited amount of energy has been generated in this country over the past 20 years without any serious incident and without any of the grim forecasts of the opposition having materialised, although there was a serious accident to a Pressurised Water Reactor (PWR) at Three Mile Island, Harrisburg, Pennsylvania in March 1979 that gave fresh ammunition for the anti-nuclear lobby. Even at Three Mile Island, however, where the safety precautions were criticised, only a few workers were exposed to more than the permitted quarterly dosage of radiation and were promptly sent home, whilst outside the plant perimeter, the maximum dose that could have been acquired by any local inhabitant was about 85 millirems, compared with 100 millirems from a single x-ray examination and 100 millirems from a year's exposure to cosmic rays, natural radioactive sources such as granite, and colour television sets.

The more extensive use of nuclear energy should be particularly attractive to chemists, because to us, converting oil and coal to carbon dioxide is a waste of hydrocarbons which could be used for making so many useful and valuable chemicals. And in any event the importance of taking stringent safety precautions should be just as much a part of a chemist's or chemical worker's philosophy as of a worker in the nuclear power industry. So I would support the present policy of building several different types of nuclear reactors before a final decision is made about the type to be chosen for large-scale development in future, but it is essential that more attention be given to the development of the fast breeder reactor as this enables a 50-fold increase to be made in the amount of energy generated from uranium; and it can utilise the plutonium recovered from the spent fuel of thermal reactors. The risks involved in the transportation of this spent fuel to reprocessing plants such as those at Windscale or La Hague, the recovery of the plutonium, the disposal of the neptunium, americium, curium and other radioactive wastes with long half-lives, and the fear that small amounts of plutonium may be stolen in transit by terrorists have become strong arguments for the anti-nuclear activists. Even though the formation of vitrified blocks for sealing up the long-lived isotopes would seem to be better than any other method so far devised for limiting the spread of radiation, what to do with these blocks during the several thousand years required for the radio-activity to decay is still highly uncertain. Unfortunately it may be 20 or 30 years before the feasibility of nuclear fusion is demonstrated, but once it can be shown that the fusion of deuterium and tritium produces energy that can be recovered and utilised, then it will no longer be necessary to mine uranium, reserves of which will be approaching exhaustion by that time. Fortunately sources of deuterium and tritium are readily available and abundant. Sea water contains 35g of deuterium per m^3 and 500g of deuterium can produce energy equivalent to 1500 tonnes of coal. Nuclear fusion could well be the ultimate answer for the future.

Renewable Forms of Energy

Of course, the anti-nuclear lobby would advocate by-passing all forms of nuclear energy completely and developing some of the renewable forms of energy without further delay. Solar energy is one of the most attractive, creating no environmental hazards, but in Europe north of the Alps sunshine is so limited and sporadic as to be suitable only for intermittent use and quite inadequate for industrial use, though it certainly has a useful role for domestic heating, and one large hospital in south-west England has installed solar panels with an area of 40,000 square feet. Yet, because the amount of solar energy reaching the earth's surface in a week is equivalent to the planet's total proved energy sources, new ideas are being explored. These include the electrolysis of water to generate hydrogen, improving the energy return from photosynthesis, simulating the process of photosynthesis itself as in Sir George Porter's work at the Royal Institution, sending satellite solar power stations into orbit to collect solar energy for 24 hours a day and transmitting the energy to the earth by micro-waves and erecting a cluster of large parabolic mirrors to focus the sun's rays on an industrial furnace as at Montlouis in the Pyrénees Orientales. But the realisation of most of these schemes lies well into the future and will require large sums of money and wide international cooperation.

Windpower generators are another attractive possibility which nevertheless has limitations. Improved machines have been developed that could make a useful contribution to local farming communities, and in the Orkneys one is being built that will make a substantial contribution to the island's electricity supply in about three years' time, and there will be no damage to the environment.

But more than this is expected from the third of the renewable forms of energy - tidal energy, which has long been an accepted source at La Rance in Brittany and in the USSR. In this country, harnessing tides would present serious environmental problems, since certain sites would conflict with other possibilities such as potential fresh water reservoirs, power stations, and interference with drainage, sewage and navigation. The most practicable scheme is the proposed Severn

Estuary Barrage which would save some 5m.t.c.e. of fossil fuels per year, would provide some environmental benefits to offset some of the disadvantages, but would create some long-term ecological effects. But the cost is enormous and it may be 10 to 15 years after a favourable decision has been taken before any electricity would emerge from the scheme.

The apparently related proposal to extract energy from waves is in reality very different and particularly attractive to a nation surrounded by oceans with waves reaching its shores containing large quantities of accumulated potential energy. Removing this energy may in due course perhaps alter the shore lines by depositing sand and silt in different places from the original ones and may result in ecological changes on shore. And of course methods of recovering this energy must not interfere with shipping, harbours, the discharge of rivers or effluent from sewerage works, with nuclear power stations or other shore-based operations. The preferred site for wave machines is somewhere off the Hebrides, and several types are at present undergoing tests - Stephen Salter's "nodding ducks", Sir Christopher Cockerell's articulated rafts and Michael French's flexible "stockings", and the electricity produced by them would be transmitted by cable to the mainland, or used to produce hydrogen and oxygen in an electrolytic cell.

Geothermal energy is another attractive possibility for this country that appears to have no adverse effect on the environment, but is likely to be expensive. The extraction of warm underground water will possibly only be suitable for providing domestic heating in certain areas of the country, but the creation of steam by injecting water into hot, dry impervious rocks after these have been artificially fractured has far greater potential, the granite rocks in the south-west of England alone, it is estimated, containing 6000m.t.c.e. and yielding electricity at 2-3 p. per kw.

Biomass

Another interesting renewable source of energy is "biomass", renewable because it is based on the conversion into liquid or gaseous fuels of organic wastes from farms, of domestic refuse and sewage, of general industrial wastes and, particularly in the tropics, of specially grown "fuel crops". Such materials could make a substantial contribution to our energy requirements, even in the UK, where an optimistic forecast of as much as 50 m.t.c.e. per annum has been made, whilst Brazil hopes soon to replace up to 25% of its oil imports by alcohol produced from agricultural crops, and similar developments are taking place in other countries. A variety of processes is available, such as aerobic fermentation with yeasts, anaerobic fermentation with bacteria, pyrolysis and, perhaps most promising of all, steam gasification. The products would depend on the materials available and the process chosen - ethanol, methanol, methane or liquid hydrocarbons.

The Chemical Industry's Requirements

Where then does the future of the chemical industry lie, and what feedstock will it use? Petroleum will last some time yet, but will cost more and more as time goes on, and will become less and less attractive. Is the answer to be coal or nuclear energy, "biomass" or some other form of renewable energy? When asked this question at a recent conference, Sir Hermann Bondi's reply was "we must make use of <u>everything</u>, then we might just get by". Clearly the environmental hazards associated with different sources of energy will be an important factor in determining which sources will be chosen, and of course the chemical industry will be anxious to ensure that enough coal is going to be reserved for use as a raw material and not all consumed to provide energy; in addition it should give its support to any new form of energy such as "biomass", that would provide additional sources of the raw materials it needs.

Let us return to the more immediate future, say the next two decades, and look at two other sources of raw materials for chemicals which are likely to become available to the industry when coal takes its former place as the main source of energy. North Sea gas is not likely to be exhausted for some time yet, but the British Gas Corporation is already planning how the gasification of coal can eventually produce substitute natural gas for distribution through the existing pipe-line system to the homes and factories that at present use natural gas for heat or power. It is impossible to say at this moment what processes will ultimately be used to achieve this objective, but the proposed process at the moment is the Lurgi Slagging Gasifier, developed in the USA, some 20 of which would be needed each requiring about 200 acres of land. The creation and subsequent operation of each plant would cause considerable interference with the neighbourhood, it would require a substantial workforce and could seriously pollute the atmosphere and create problems of how best to dispose of solid wastes. However it is planned to operate the plants in such a way that most of the sulphur would be recovered either as elemental sulphur or sulphuric acid and the liquid effluents processed so that valuable by-products such as ammonium salts, phenols and hydrocarbons would be recovered.

The other development that should be mentioned is the production of liquid fuels from coal, since coal is likely to become the major source of liquid fuel for transport, and again, the by-products would be of great importance to the chemical industry. There are two routes by which coal can be converted into liquid fuels. The first, known as the synthesis route, involves gasification of the coal to produce raw synthesis gas, which is mainly carbon monoxide and hydrogen, and reaction in presence of catalysts to synthesise the organic fuel molecules, which vary according to the method used, the best known being the Fisher-Tropsch process. By the second route, the coal is degraded either by heat alone or in presence of an organic solvent, and hydrogen is added during or after degradation, either as hydrogen gas or as a hydrogen donor in presence of a catalyst. The crude product may then be subjected to catalytic hydro-cracking similar to that used in the refining of crude natural oil. Three different coal hydrogenation processes have been developed in the USA and more are under development, whilst in Britain two variants of the degradation/hydrogenation route are being investigated by the National Coal Board at Stoke Orchard, one comprising liquid extraction with a process-derived solvent which behaves as a hydrogen donor, and the other extraction by a gaseous hydrocarbon at supercritical temperature and pressure. The liquid fuels produced by these processes are different from petroleum-derived products and unfortunately are more active physiologically, even though the benzo(a)pyrene content is not much higher; they will also have a much higher nitrogen content than the petroleum products, much more work will have to be done to ensure that fuels produced by these processes do not possess carcinogenic activity.

So far as this country is concerned therefore, oil will gradually cease to be a major source of energy or petrochemicals during the next two decades and during this time, coal will take the place of oil for both purposes, and some additional energy will continue to be derived from nuclear power stations. The amount of nuclear energy will not increase substantially, however, until agreement is reached internationally about the disposal of radioactive waste with a long half-life. Unfortunately there is as yet no evidence that nuclear fusion can replace nuclear fission. Beyond the next two decades it is impossible to predict the future. Some of the proposals now under examination seem too grandiose even to be capable of realisation, and many will require a degree of international cooperation that at the present juncture seems highly unlikely to be attained. Others are only capable of producing a fraction of the energy that will be needed, but may nevertheless have to be used. Fortunately, by the year 2000 the results of much research now in progress may be available that should enable some decisions to be made more realistically than is possible at present. We should know more about the feasibility, for example, of using nuclear fusion as a source of energy, of simulating photosynthesis, and of extracting energy from waves and hot rocks and from the waste materials that will be produced in ever-increasing amounts from an ever-increasing population requiring ever-increasing amounts of energy.